she has given us a spirited and beautifully written account of what dreams the soil can hold."

— **JANE BROX**

AUTHOR OF *CLEARING LAND: LEGACIES OF THE AMERICAN FARM*

"From graham crackers to heirloom carrots, Kelley offers an engaging and thorough dive into the long, strange history of American dreaming and eating. Essential reading for anyone wondering not just where their food comes from, but why."

— **KATE DALOZ**

AUTHOR OF *WE ARE AS GODS: BACK TO THE LAND IN THE 1970s ON THE QUEST FOR A NEW AMERICA*

"This book tastes so good—I ate the whole thing raw."

— **MARK SUNDEEN**

AUTHOR OF *THE UNSETTLERS: IN SEARCH OF THE GOOD LIFE IN TODAY'S AMERICA*

"*Foodtopia* is an engaging, informative, and inspiring journey, exploring the deep political and ecological values that motivate alternative agriculture. Margot Anne Kelley reveals the historical continuity that weaves together healthy food, community agriculture, and racial justice. In so doing, she offers an exciting and hopeful vision of a great American tradition."

— **MITCHELL THOMASHOW**

AUTHOR OF *TO KNOW THE WORLD: A NEW VISION FOR ENVIRONMENTAL LEARNING*

"Margot Anne Kelley's *Foodtopia* is a marvelous book tracking the history of five waves of utopian back-to-the-landers in America. From Brook Farm to the Diggers, from Henry David Thoreau to Farm Aid to Alice Waters, and on to the millennial farmers bringing queer and BIPOC perspectives to food production, this book is a gorgeous cornucopia. And it makes clear that these movements are not an escape from the struggles of race, class, and privilege. Rather they are ground on which to think anew about cultivating both healthy food and a just food system. A stellar and timely book!"

— **ALISON HAWTHORNE DEMING**

AUTHOR OF *A WOVEN WORLD: ON FASHION, FISHERMEN, AND THE SARDINE DRESS*

"The impulse to escape the city and reconnect with a more authentic world through food and farming has been a perennial part of the American experiment, but as *Foodtopia* shows in clear and compelling prose, it's much more than that: By challenging the centralized dogmas of mainstream culture through a kind of agrarian performance art, these troops of back-to-the-landers have not just held a mirror up to American society, but given us a lodestar so that we might re-find our way, again and again."

— **ROWAN JACOBSEN**

AUTHOR OF *AMERICAN TERROIR: SAVORING THE FLAVORS OF OUR WOODS, WATERS, AND FIELDS*

"To tackle the dire ecological and social challenges before us, we need a profound shift of direction, away from the runaway corporate-dominated economy based on high-tech and urbanization. We need to rebuild local agrarian economies, to encourage a small farm renaissance that creates a healthier balance between rural and urban. *Foodtopia* expertly documents the past movements that already clearly saw the need for this shift and whose experiences will help light the way for today's new agrarians."

— **HELENA NORBERG-HODGE**

AUTHOR OF *LOCAL IS OUR FUTURE: STEPS TO AN ECONOMICS OF HAPPINESS*

"Having been a back-to-the lander in Maine in the 1970s, I was so pleased to read Margot Anne Kelley's history of those days and her interesting insights into those Americans charting a sustainable rural life today. She does much of this through the lens of our changing relationship to food and farming. Kelley makes the strong case that changing our food system is central to everything from the quality and availability of the food we eat to the fundamentals of our very democracy. I agree."

— CONGRESSWOMAN CHELLIE PINGREE

"Margot Anne Kelley elegantly unearths the deep roots of today's back-to-the-land movement, linking Henry David Thoreau's nineteenth-century essays to the twenty-first-century struggle for food justice. *Foodtopia* shows that the desire to leave the city, grow one's own food, and live more simply is almost as American an impulse as building highways and skyscrapers."

— JONATHAN KAUFFMAN
AUTHOR OF *HIPPIE FOOD: HOW BACK-TO-THE-LANDERS,
LONGHAIRS, AND REVOLUTIONARIES CHANGED THE WAY WE EAT*

"It has been said that no reality was ever created by realists, and the utopian movements that Margot Anne Kelley explores in her joyful book all took that to heart. She observes that they share two great things: a love of good food and a commitment to a new social order. And, indeed, today the food movement in its many aspects is at the forefront of driving positive social change. Her book is full of good stories, well told, and highly motivational. Highly recommended."

— GUS SPETH
AUTHOR OF *AMERICA THE POSSIBLE: MANIFESTO
FOR A NEW ECONOMY*

FOODTOPIA

FOODTOPIA

COMMUNITIES IN PURSUIT OF PEACE,
LOVE & HOMEGROWN FOOD

MARGOT ANNE KELLEY

Boston

⫸ GODINE ⫷

2022

Published in 2022 by GODINE
Boston, Massachusetts

LIBRARY OF CONGRESS CATALOGING-IN-PUBLICATION DATA
Names: Kelley, Margot Anne, author.
Title: Foodtopia : communities in pursuit of peace, love & homegrown food
    / Margot Anne Kelley.
Description: [Boston, Massachusetts] : Godine, [2022] | Includes
    bibliographical references and index. |
Identifiers: LCCN 2021058315 (print) | LCCN 2021058316 (ebook) | ISBN
    9781567927306 (hardback) | ISBN 9781567927313 (ebook)
Subjects: LCSH: Agriculture, Cooperative--United States--History. |
    Agriculture--Social aspects--United States--History. | Utopias--United
    States--History. | Food habits--Social aspects--United States--History.
    | Food supply--Social aspects--United States--History. | Sustainable
    agriculture--United States--History. | Urban-rural migration--United
    States--History.
Classification: LCC HD1491.U5 K45 2022 (print) | LCC HD1491.U5 (ebook) |
    DDC 334/.6830973--dc23/eng/20220201
LC record available at https://lccn.loc.gov/2021058315
LC ebook record available at https://lccn.loc.gov/2021058316

First Printing, 2022
Printed in the United States of America

For E.P.B.

Let America be the dream the dreamers dreamed.
—LANGSTON HUGHES

To be realists about ourselves, we must also be utopians,
because that is the truth of our condition and aspiration.
—JEDIDIAH PURDY

There's nothing more political than food. Who eats?
Who doesn't? Why do people cook what they cook? It is
always the end or a part of a long story, often a painful one.
—ANTHONY BOURDAIN

Our most profound engagement with the
natural world happens on our plates.
—MICHAEL POLLAN

❧ CONTENTS ❧

INTRODUCTION   xiii

Chapter One
MILLENNIALS MOVE TO THE LAND
3

Chapter Two
HENRY DAVID THOREAU'S SEARCH FOR FREEDOM
47

Chapter Three
SECULAR UTOPIAS OF THE 1840s
78

Chapter Four
GOING BACK TO THE LAND DURING AMERICA'S FIRST GILDED AGE
107

Chapter Five
SCOTT AND HELEN NEARING LIVE THE GOOD LIFE
129

Chapter Six
FOOD FIGHTS
165

Chapter Seven
HIPPIE FOOD
189

Chapter Eight
TRANSFORMING THE FOOD SYSTEM FROM SEED TO TABLE
217

Chapter Nine
FREE THE PEOPLE, FREE THE LAND
251

Acknowledgments   283
Notes   285
Bibliography   295

INTRODUCTION

I N THE EARLY 2000S, when I was still an urbanite, I taught photography at the Art Institute of Boston. The college was housed
in a rehabbed parking garage, speed bumps still evident on the
floors. Everything about both the building and its surroundings
was decidedly urban and decidedly *not* gentrified. It sat steps
from Kenmore Square, which meant AIB students and faculty
could easily enjoy the conveniences of a city—though the flip
side included learning how to avoid the pickpockets who flocked
to Fenway Park when the Red Sox were playing at home. The
school's only greenery was a garden in the bed of a pickup truck
that one of the librarians parked out front, on Beacon Street, on
warm days. As might be expected of budding artists, the students
were regulars at the galleries, especially the ones serving free food
and wine. Regardless of where they were from, by the end of their
freshman year, most of the students were city-savvy and eager to
appreciate what Boston had on offer.

Yet every semester, at least one—and often several—of those
young hipsters worked on projects about small-scale farming or
sustainable living. We faculty chalked up the proliferation partly
to Michael Pollan's *The Omnivore's Dilemma* (2006) and Barbara
Kingsolver's *Animal, Vegetable, Miracle* (2007). Pollan's book al-

most instantly helped to broaden and deepen the national conversation about what Americans eat. Many of my students were astonished—and appalled—to learn how industrial agriculture works and the ways the US government puts its thumb on the scale in favor of huge, monoculture farms. Kingsolver's book made the prospect of opting out of the industrial model and eating a local diet seem within reach: By chronicling her family's year of eating only local food, she tacitly assured readers that Americans who weren't uber-wealthy could do it.

Like Kingsolver, many students set out in search of something better than the status quo. Quite a few found it on small farms throughout the region, some traveling as far north as New Hampshire and southern Maine and others to Cape Cod, exploring and documenting places run by folks just a few years older than themselves. One young woman set herself the task, beginning in late January, when the term started, of eating only locally grown food for a semester. Another made wryly elegant photographs to kindle conversations about GMO crops. One spent a semester living in an off-the-grid collective in Waldo County, Maine, striving to be self-sufficient and carbon neutral.

I may have seen more than my fair share of food and farm projects, as students knew those issues mattered to me. In 1999, my husband and I had bought an old farmhouse on a peninsula in Maine; I spent summers there and Rob spent whatever time he could carve out of his hectic schedule as a technology consultant. Our yard was mostly filled by a tangled thatch of blackberry and a dark stand of spruce trees. On one of the few sunny patches of lawn, we put in a small raised bed so I could grow vegetables. Thanks to implausibly perfect weather the first year, we enjoyed a harvest of tomatoes, peppers, salad greens, kale, and herbs; I was hooked.

The following summers, I chipped away at the blackberry thatch so we could add more vegetable space. Soon, I grew enough food that late-summer meals were dictated primarily by

what was ripe in the yard. I even put up some things for winter, which made me ecstatic. That may sound hyperbolic, but it's
true; I'd spent practically my whole life in school, having gone
straight from student to professor. Such a visceral relationship
with the natural world was new for me, a grimy, primal pleasure.
Slowly we shifted our garden and diet toward plants that fare well
in Maine. Out went the eggplant, in came potatoes.

When I learned about heirloom vegetables, I began growing
mostly these old cultivars. I loved learning their histories and often
loved their distinct flavors. At a lecture I attended about the loss of
culinary diversity in the United States, the ecologist Gary Nabhan
pointed out that if people don't eat regionally specific cultivars,
farmers stop growing them and they become extinct. This wasn't
simply a culinary concern, he said; it was also a climate issue. As
the planet warms and weather patterns change, having a lot of
plant-breeding materials will better enable farmers and scientists to
create resilient new cultivars. His "eat it to save it" ethic made me
even more eager to grow heirlooms; I sought out seeds for many of
New England's endangered fruits and vegetables, supplementing
those with interesting old breeds from other places.

Of course I began growing old-fashioned tomato cultivars, but
it was heirloom beans that provided the great surprise. Unlike
the half dozen or so in the grocery store, these old cultivars, with
names like Vermont Cranberry, Henderson's Black Valentine,
Brown Dutch, Tongues of Fire, Ireland Creek Annie, and Lazy
Housewife, are wildly diverse. The sizes, colors, skin thickness,
texture, and flavors of the many strains differ considerably.

Like my students' interest in farming and other simple living
experiments, my newfound passion for growing food wasn't unusual; we were part of a shift in the zeitgeist. The desire to eat in
a better manner—better for both people and the environment—
was prompting many young Americans to start farming and inspiring others to create gardens, live more lightly, and eat more
wisely. Much as I loved teaching, I began dreaming of a simpler

life more connected to the soil, one with Black Oxford apples and Bumblebee beans and a pollinator garden. But Rob's work as a computer consultant required him to travel a lot—often he was away the entire workweek and home only on weekends—which meant he had to have ready access to a large airport. When he could work from home, he needed reliable, ample broadband service, which the St. George Peninsula lacked.

By the end of the early 2000s, internet access had improved in parts of St. George. Around that same time, a brother-in-law of ours died and my father suffered a serious stroke. Facing those reminders of life's unpredictability, we decided not to wait for some perfect moment to revise our lives. We sold our home in Massachusetts, moved our belongings into the barn of the Maine farmhouse, and hired a local construction company to build us a solar-powered house a mile down the peninsula. As our desire for an energy-efficient place and our utter dependence on the internet suggest, Rob and I weren't leaving the contemporary world behind; we were just inching toward a more grounded life.

In April 2010, just a few weeks before the end of my final semester at AIB, my teaching assistant, Tyler Sage, and I sat outside drinking coffee and talking about his postgraduation plans. The air was filled with fumes from cars and buses and the squawk of brakes as trains paused to release queues of students and workers, but the commotion didn't derail the conversation. Tyler had had an epiphany, knew what he wanted to do: He wanted to run a small, diversified farm. Thrilled, I began suggesting books, experiences, people who might be able to help him get started. But he wasn't asking for that kind of help; he needed advice about how to talk to his parents. Apparently, the only thing more difficult than telling parents you want to be an artist when you grow up is telling them you want to be a farmer. Though his parents were dubious, Tyler was far from alone in trading a city job for life on a farm. So many young adults from non-farm families were interested in learning about small-scale agriculture that farm internships and apprenticeships became highly competitive. As Severine von Tscharner Fleming, one of the founders of the National Young Farmers Coalition, wrote in an article published the following spring, "If you talk to any really good farmer, they'll tell you that they've had a doubling and tripling of their applicant pool over the last few years."[1]

Despite that competitive environment, Tyler quickly landed an apprenticeship. As he began his journey toward farming, I began mine toward living more sustainably. Our new house wasn't ready but the garden beds were, and I had more than twice as much growing space as at the old one. *I could try anything!* One of the seed companies I ordered from was offering garden-sized quantities of a wheat called Black Winter Emmer, and I planted a whole bed of it. Because I'd never seen anyone plant wheat in real life, I resorted to mimicking Jean-François Millet's *The Sower*, and threw the two pounds of grain in expansive arcs. The result was a wildly uneven patch. But two pounds, it turns out, was ten times as much seed as I should have used, so by the beginning of July we had a tall stand of pertly green grass. By August, we had anthem-worthy amber waves of grain. The wheat was so beautiful that people walking down our road often came into the yard to admire it up close.

All summer I gushed about how easy wheat-growing was. Whereas the other beds required constant tending, this one just demanded I pull the vetch when it began girdling the stalks. Then it came time to harvest. One hundred square feet is too small an area to use anything but a scythe or trimming shears to cut plants. We laid the wheat on old bedsheets in the spare room to finish drying and then trimmed the heads—one by one. These we put into a faded pillowcase, which we smacked against the garage wall to separate the wheat from the chaff (yes, literally). Beating the wheat was absurdly fun, a small compensation for the fact that our stand yielded just a few handfuls of wheat berries, as I'd made the mistake of planting winter wheat in the spring. The yield was so ridiculously small we didn't even mill it. I began to understand why ours was the only wheat patch around.

Fast-forward five years to our second try. From the first foray, I knew to plant spring wheat and to do so in evenly spaced rows. I also knew better what to expect for yield because in the intervening years, more seed companies had started selling grain to home gardeners and making information about small-scale production available. I'd discovered an attachment to turn a KitchenAid mixer into a grain mill, which would be much easier than using our old-fashioned hand-cranked mill. Still, harvest 2.0 yielded only enough wheat to make four loaves of bread.

Some of my neighbors chuckled that we ceded space to a crop with such a low yield, but—at least this second time—we knew what we were in for. Sure, we welcome large harvests, and we had expected more than four loaves' worth, but we're growing these heirlooms not only to avoid industrial foods and reduce our carbon footprint, but also to help keep them around. I want this wheat, Red Fife, to remain in production. On its website, the heirloom grain company Anson Mills describes bread made with Red Fife with a sommelier-like flair, describing it as having "profoundly herbaceous and nutty fresh wheat flavors, a moist, satisfying crumb, and a lovely crust with deep, toasty caramel notes." Such great taste should be reason enough to keep it in production, yet it almost became extinct.

During the nineteenth century, Red Fife was one of the most commonly grown wheats in Canada and the United States, but it disappeared early in the twentieth, displaced by a strain that matured more quickly. Except in conservation gardens and seed banks, Red Fife was ignored for most of the century. Then, in the late 1980s, beginning with just a single pound of wheat, Canadian growers worked to bring it back. In 2004, Canada featured Red Fife at Terra Madre, the international Slow Food Celebration of important endangered foods. Since then, it's attracted a following, thanks to concerns about the nutritional paucity of white flour, worries over genetically modified foods, and an increase in gluten sensitivities.

I realize planting two pounds of Red Fife won't ensure its survival, but I hope it sends a small message to seed companies that a market exists for these older, healthier strains. Plus, when we make sourdough bread with wheat from our garden and starter Rob wild-fermented, we savor bread that's truly *homemade*, the yeast derived from the bloom of berries we grew, along with others released by nearby trees and weeds. Romantic? Maybe. Awesome? Absolutely.

Some neighbors laughed at my wheat, but others pulled into our gravel driveway to reminisce about their own long-ago efforts growing grain. Soft-spoken, older men whom I'd bought firewood from or met at town meetings stopped by, lowered the driver's side window of their pickup or car, sometimes even turned off the engine. With pleasure, they regaled me with stories from their idealistic, sometimes decadent youth: A fair number of people moved to Maine as part of the hippie-era, utopian, back-to-the-land movement.

During the late 1960s and well into the 1970s, roughly a million young adults fled cities and suburbs in search of a freer, happier, more authentic lifestyle in rural communes and cabins. Eventually, a majority returned to the mainstream they'd spurned, but enough of those who had decamped to rural areas across America remained that they transformed local cultures. In my adopted home state, "the movement somehow worked its way into the tightly woven fabric of the state's culture," writes Joseph Conway. "Like moose, lighthouses and lobster, the back-to-the-land movement is, for better or worse, Maine."[2]

Rob and I aren't part of their generation. When those counterculturalists were embarking on their grown-up lives, I was a proto-hip little kid listening to Marlo Thomas's children's album, *Free to Be . . . You and Me*, with its promises of an idyllic, gender-neutral, socially tolerant world "where the river runs free, through a green country, to a shining sea." Nor are we back-to-the-landers, though we have in common many of their values and a bit of their DIY ethos. We rely on solar power, grow a lot of food, make cheeses and soap, use a catchment system to collect rain to water the garden.

The hippie-era homesteaders were way more hardcore. Many not only grew their food, but also constructed their cabins or yurts, used natural remedies, and had their babies at home. Most steered clear of modern conveniences for ideological reasons as well as economic ones. Objecting to the military-industrial complex and fearing nuclear power, for example, prompted many to live without electricity. Most tried to limit their interactions with capitalism, preferring making do, sharing, and bartering to participating in consumer culture.

As the former grain-growers told their stories, I was intrigued by how much they had in common with new farmers like Tyler. And as more young farmers started selling their wares at area farmers' markets, I noticed the vendors were almost all Baby Boomers or Millennials: Plenty of people of my vintage, Gen Xers, were enthusiastic customers, but few were merchants. Implausible though it seemed, it looked like the impulse to live lightly and with greater self-sufficiency simply skipped a generation—mine. Curious (and a little jealous), I began investigating why that might be.

As it turns out, this impulse doesn't skip a generation, though it does wax and wane. It's been doing so in the United States for almost two centuries—since the nation's population started to shift from mostly rural to mostly urban. The first back-to-the-land movement occurred during the 1840s; in fact, that era was regarded as the golden age of back-to-the-land experiments for more than a century—until the hippies decamped. The second large back-to-the-land push occurred around 1900 and a third took place in the 1930s. Hippies headed for the countryside at the end of the 1960s and into the seventies. And young people from non-farm families who started going to the land to farm at small-scale during the late 2000s comprise the fifth of these back-to-the-land movements.

Perhaps surprisingly, given the notion that rural America is inherently conservative, factors leading these folks to try to live lightly on the land are almost all socially, spiritually, or politically progressive. Yearnings for racial and gender equality, for peace, for a less consumerist lifestyle, for simplicity, for a healthy diet, and for a sustaining connection to the natural world animate the participants. All of that sounded great to me, which made it even more puzzling that my generation didn't head for the fields. Why, I wanted to know, did these movements occur only in fits and starts?

One reason was obvious: A life lived close to the land requires hard work, often grindingly hard work. I've talked to farmers who haven't been off-farm overnight in a year because they don't have anyone to ask to tend their livestock, and to erstwhile homesteaders who moved to the country with fiery ideals but no idea what woods burn well. With so much unfamiliar and physically difficult work involved in self-sufficient rural living, sizable migrations occur only when there's a groundswell of urban dissatisfaction. What leads to and connects these movements is the participants' shared belief that they can create an alternative social order and their sense that they must, because something about mainstream culture strikes them as horribly wrong.

In short, they're utopians.

By that, I don't mean starry-eyed idealists pursuing impossible projects. I mean folks who have a particular kind of double vision, who see the existing world clearly and envision alternatives they're willing to work to bring into being. The utopian studies scholar Lyman Tower Sargent calls this "social dreaming," a phrase that nicely describes much of what I'm focusing on here. Social dreaming involves asking and answering questions about what could constitute a better alternative to current living conditions, and then actually pursuing that way of life. To be sure, many social dreamers work within society for social or political change. But *performing* an alternative, demonstrating it, distinguishes utopians from others who also seek change. The utopians create microcosms not only for their own benefit, but also for the sake of the rest of us: They lead by example. Which is not to say they're preachy or sanctimonious—though some most certainly were.

Utopians often get a bad rap for trying to create ideal alternatives, mocked for being naive or foolish or unrealistic. Then, if they can't maintain a perfect community in an imperfect world, they get derided for failing. And most utopian experiments do fail, succumbing to in-fighting, or losing their cohesion when a leader dies, or running out of money, or, or, or. When Fruitlands, a utopian experiment led by Amos Bronson Alcott, failed after less than a year, his daughter Louisa May griped that "the world was not ready for Utopia yet, and those who attempted to found it only got laughed at for their pains . . . the failure of an ideal, no matter how humane or noble, is harder for the world to forgive and forget than bank robbery or the grand swindles of corrupt politicians." She was right; mainstreamers often seem to relish their downfall.[3]

But if we look to the origin of the word *utopia*, we find something neither particularly foolish nor laughable. In 1516, Sir Thomas More coined it for his book *De optimo rei publicae statu deque nova insula Utopia*, better known simply as *Utopia*. He was punning on the Greek words for "no place" (*ou-topos*) and "good place" (*eu-topos*), seeming to imply that this ideal can't exist in the real world, and he reinforced that belief by naming his narrator Hythloday, which means "nonsense peddler." The wordplay helped provide More with plausible deniability—a way to say *Hey, it's just fiction!*—if he faced censure for criticizing aspects of England's political and social life.

Despite broadly hinting that Utopia existed only in his imagination, More's vision was inspired by accounts the explorer Amerigo Vespucci had published a few years earlier. Traveling in the Americas, observing Indigenous Peoples there, Vespucci wrote that the first ones he encountered were self-ruled: They "have neither king nor lord, nor do they obey anyone, but live in freedom." They lived communally, many to a household. And they didn't use money. They "have no commerce," he wrote, "and neither buy nor sell. In conclusion, they live, and are content with what nature has given them."[4]

More's Utopians, likewise, have no need for money and no interest in social rank or hierarchy. They live in an egalitarian and communal fashion, taking what they need from local warehouses, eating together, and generally working no more than six hours a day. More suggests this way of life is much better than that in Europe; in Utopia the problem of poverty has been solved and humans are guided more by their better angels than their devilish counterparts. Not that Utopia is perfect: Every household has slaves, for example, privacy is nonexistent, and the ability to travel is tightly constrained.

Still, Utopians were far better off than the English peasants. For centuries, the peasants had been able to grow and glean crops, gather firewood, and hunt and fish on privately owned "common land." But starting in the fourteenth century, access was becoming curtailed through the practice of enclosure. The commons were taken by wealthier people, claimed as private property, and used for pasturing sheep. Without access to the commons, the poor struggled to feed themselves. Few could hire out, as the land wasn't being cultivated. Some resorted to stealing—a crime punishable by death.

In More's Utopia, in contrast, land access and farming practices are equitable. Each citizen works as a farmer for two years during early adulthood. The farmers live in large, communal households, and at the end of each harvest, half the household returns to the city and is replaced by a new group. In this way, agricultural knowledge is passed among all, and no one is required to perform this arduous work for too long. After their farming stints, citizens join the family business or pursue a strong interest.

But this widespread acumen ensures that ample skilled labor is available at harvest. When crops are ready, the farmers let municipal officials know how much help they need and "the crowd of harvesters, coming promptly at the appointed time, dispatch the whole task of harvesting almost in a single day of fine weather."[5]

Of course, not every real-life utopian goes back to the land, nor is every back-to-the-lander a utopian. This book is about the five times that large numbers of utopians in the United States enact their ideals by doing so—in the 1840s, around 1900, in the 1930s, in the 1960s and 1970s, and in the 2010s. Like all other utopian efforts, theirs were reactions to specific conditions they wanted to avoid or fix. And even though much has changed in America since the 1840s, the back-to-the-land groups I'm highlighting have in common many core beliefs. All of them valued the concept of freedom that propelled the founding of America, and all of them feared such freedom had become more difficult to attain. All believed inequality hampers society. All thought Americans lost an essential connection to the natural world. And all thought food is often debased, which undermines the one assured connection people have with nature.

The idea of freedom the back-to-the-landers cherish was radical when America's founders declared it. According to the constitutional law scholar Jedidiah Purdy, its "most explosive element" was the claim that people are "naturally free" and governments are meant to "protect their inborn rights and interests." If governments failed in that charge, then people "could take down and rebuild those governments as they saw fit." This vision wed individual, natural freedom to a broader national identity, a combination that made eminent sense at the outset, when colonists sought to liberate themselves from British rule.[6]

Over time, its limits became impossible to ignore, most obviously in the inherent contradiction of allowing slavery in a nation grounded in natural freedom. As industrialization made society more consumerist, other social and economic inequalities also became more apparent. In successive generations, people often sought to expand access to the freedom the founders accorded to white, property-owning men. Thus, they tried, as Purdy writes, to "bridge the gap between the utopian demand for a world as capacious as our dreams and the persistent reality of limits" by cultivating a "tradition of vision and experimentation." The utopians in this book are among those who sought to create a world as capacious as their dreams; they did so not for themselves alone, but for everyone. Some rejected the nuclear family, believing it subjugated women. Others rejected private property, feeling it ensured economic inequality. Some worked to abolish slavery or to broaden civil rights, hoping to bring about equality among the races.[7]

Their emphasis on equality extended beyond the human realm. Rather than dividing the world into people and everything else (as do many philosophies and belief systems), these utopians saw non–human beings as subjects in their own right. Bronson Alcott, the transcendentalist leader of Fruitlands, for example, believed animals should not be eaten or exploited because people were supposed to help guarantee animals' happiness. Alcott's friend and fellow transcendentalist Henry David Thoreau took this notion even further. When someone asked if he was lonesome living by himself at Walden Pond, Thoreau replied: "I am no more lonely than a single mullein or dandelion in a pasture, or a bean leaf, or sorrel, or a horse-fly, or a bumblebee . . . no more lonely than the Mill Brook, or a weathercock, or the north star, or the south wind, or an April shower, or a January thaw, or the first spider in a new house." Not only did he place these various entities on the same level, rather than in a hierarchy, he also intimated that they weren't as separate from each other as they seemed. He went so far as to describe himself as "partly leaves and vegetable mould."[8]

Wendell Berry, an award-winning poet, essayist, and novelist who is also a well-known farmer, environmentalist, and philosophical provocateur, believes this awareness of deep interconnection between people and land used to be intrinsic to our national identity, but that we've lost our consciousness of it. Americans, he says, have forgotten "that our land passes in and out of our bodies just as our bodies pass in and out of our land; as we and our land are part of one another, so all who are living as neighbors here, human and plant and animal, are part of one another." Unlike many of us, the back-to-the-landers discussed in this book never lost that sense of profound reciprocity or the belief that equality is necessary to freedom.[9]

Perhaps most surprising among the different groups' common concerns is good food. To be sure, food often plays an important role in utopian experiments, as it can reflect and reinforce group values. But it plays an especially important one to these back-to-the-landers. For them, "good" is tied as much, or more, to ethics as it is to flavor. Those ethics infused both their food choices and their cuisines more broadly. They paid attention to what's eaten, how it's prepared, with whom and when and where it's eaten, how it's produced and distributed, and what it looks and tastes like. Their decisions weren't all the same, but all were shaped by the desire to be more self-sufficient and to evade mainstream practices they found troubling.

Crucially, they believed the correct production and consumption of food is essential to becoming and remaining free. Often, they feared mass-produced items were adulterated or contaminated. All avoided foods produced using exploitative practices that harm laborers or the environment. By growing their own food, or eating lower on the food chain or lower in the food system, or advocating for food justice, or demanding stricter food-safety measures, the utopians enacted their ideals through their meals.

Cherishing freedom, equality, and a connection to the land and to other species. Considering food important. Check, check,

check, check, and check. I share those values, as do plenty of other Gen Xers. So why didn't we decamp? experiment with utopia-making? The main reason, I think, is that although the 1980s (when the oldest Xers were becoming adults) were hardly an idyllic time for progressives, the 1990s (when most Xers came of age) were quite a bit better. The decade had a strong economy; the number of deaths from AIDS was declining; the internet expanded into something interesting to people not in DARPA; cell phones got powerful enough to be useful. In short, it wasn't a bad time to be coming of age.

In contrast, the utopian impulse I'm exploring is strongest when it's a bad time to come of age, when America's emphasis on rugged individualism is so excessive that enjoying a free and fulfilling life in the mainstream feels extra challenging. When inequality is acute, when opportunity is limited, when freedom is in short supply, large numbers of utopians turn to nature to cultivate an intimate connection to place and a commitment to community. There, they try to live outside systems they blame for making the mainstream less healthful and less joyful than it could be, especially the ones that consolidate power, reinforce hierarchy, and heighten inequality.

Foodtopia is the story of these renegades. Some are well known, such as Henry David Thoreau; others, far less so. Famous or not, they have much in common, often having learned from and been inspired by one another. In fact, from Thoreau living at the edge of Walden Pond in the 1840s to Vera Fabian and Gordon Jenkins cultivating Ten Mothers Farm today, we can trace a direct line of interpersonal influence. These utopian experimenters have daisy-chained into being an enduring counterforce to the mainstream ethos, one that's based on a love of freedom, equality, reciprocity—and good food.

FOODTOPIA

1

MILLENNIALS MOVE TO THE LAND

O N A BLUSTERY SPRING morning, I headed to the Appleton
Ridge, a beautiful, elevated stretch about an hour from my
house in Midcoast Maine. I wanted to talk with Ethan and Elizabeth Siegel, who run Heritage Home Farm, about their journey
from city life to running an heirloom, organic livestock farm.

In 2012, disenchanted with their jobs and horrified by what
they'd been learning about genetically modified foods, Ethan
and Elizabeth set out to live more lightly, closer to nature, and
in a way that could improve the health of people and planet.
They spent two years apprenticing on eight farms in four states—
Colorado, Vermont, Pennsylvania, and Maine. "We fell in love
with Maine," Elizabeth told me. They continued apprenticing on
farms throughout the state until they were able to find land to
buy. In 2015, they began to put down roots at what is now their
organically certified farm.[1]

As Elizabeth and I toured the farm, a bottle baby goat joined
us. Born two weeks early to a first-time mom who didn't take
to mothering, the kid had to be fed by bottle. When the wind
made talking outside too difficult, Elizabeth and I headed into the
greenhouse, and the kid followed, intent on staying close. Elizabeth said this was the first year they were able to start seeds in the

greenhouse; the previous two years gusts coming over the ridge had torn off the plastic sheeting. She laughed it off good-naturedly, attributing both the plastic woes and the ill-trained kid to her and Ethan's still-limited experience. This year, she said, they'd hired a professional to skin the greenhouse.

So far, they had planted seeds for twenty-four crops. Most were chosen based on their historical and cultural significance—such as the watermelon from Iraq her father and aunt ate as children and the Kabuli and golden garbanzo beans used to make hummus. But, Elizabeth emphasized, they chose them for flavor, not nostalgia, growing only heirloom foods that are as delicious as they are culturally significant. Even though she acknowledges that newer seed lines may grow faster or be more productive, her first question about a cultivar is always "What does it taste like?" In traditional dishes, she said, you can really taste the difference: "If you're growing an Italian heirloom squash and you're making ratatouille, it's going to taste more authentic compared to if you're just using the conventional or hybrid squash you can find at the store."

Many farmers think genetically modified crops are "the future," but the young farmers who've gone back to the land since the Great Recession prefer open-pollinated crops and organic methods, both for taste and for their benefits to the environment. Avoiding GMOs enables them to avoid the herbicides and pesticides that GMO crops require, which often harm soil fertility and water quality. "I hope farming in the United States will move full circle," Elizabeth told me. "I think that's what we're doing now. The food in our country got so bad that it's finally gotten to the point that young people are becoming farmers again." She attributes this, in part, to social media and the internet: Companies specializing in genetically modified products, she said, "have been doing what they've been doing since I was a child, but now we're more aware . . . It's harder now to hide things from people, especially young people. I hope things will keep moving in our direction."

I hope so, too—and we have reason for some optimism.

Most young farmers come from farm families. But starting toward the end of the 2000s and continuing into the 2010s, more than five thousand young Americans like Ethan and Elizabeth, people who aren't from farm families, became the "primary operators," as the USDA calls them, of small farms. In doing so, they became part of an influx that reversed a decades-long decline in the number of young adults entering farming. This cohort has chosen to farm for reasons that combine an interest in healthy food with environmental and social-justice ideals. The National Young Farmers Coalition believes this generation must transform America's agricultural model to make farming more socially and economically just and to improve its environmental impacts. That task would be herculean if only the five thousand primary operators were working on it. But that figure doesn't take into account the many farm employees, interns, and apprentices who were also new to farming and were embracing new methods. In 2018, Jack Algiere, the farm director for Stone Barns Center for Food and Agriculture, estimated that between three hundred fifty and five hundred new small-scale farms are established annually.[2]

As was the case in earlier back-to-the-land movements, these young people began growing food because they wanted to avoid perpetuating a problem that hampers the mainstream. They saw that the US food system contributes to many of society's problems—by abetting economic and racial inequality, by degrading the environment, and by reducing the healthfulness of foods overall. These new back-to-the-landers hope to improve the well-being of people and planet by nurturing the soil, growing plants in a healthful manner, raising animals according to their needs and wants, providing good working conditions for themselves and their employees, and giving customers a good product at a fair price.

The Millennial back-to-the-landers exhibit the same core traits as do the other, earlier utopian groups whose stories I'll also recount in these pages. They embrace the founders' notion of freedom and true equality, a sense of community and reciprocity, a desire to perform utopia, and an emphasis on good food. And like all other utopians, the Millennial back-to-the-landers are acting into being the world they want, dreaming it into existence— though by dreaming, I really mean working incredibly hard.

To put that in more concrete terms: They aren't simply shopping at farmers' markets; they're also staffing and supplying them. They aren't just emailing legislators urging them to make food access more equitable; they're growing and distributing food. They aren't waiting for Congress to raise the agricultural minimum wage; they're paying their employees well. They aren't simply hoping racism will go away; they're running antiracist enterprises and leading antiracism trainings. Many BIPOC farmers offer support to one another, and many white farmers offer allyship. This generation also performs its ideals so others can see an alternative way of living and realize that it's possible. To encourage others to sense the value of a life lived close to the land, young farmers routinely invite folks onto their farms. Visitors can go on hikes, enjoy seasonal meals, take workshops, pick up CSA shares, buy from the farm store, pick fruit, pet lambs and kids, and help with chores on community days.

Whereas earlier generations of utopian back-to-the-landers usually lived in close proximity, in communes or in clusters of small homes, this generation mostly doesn't. The earlier groups lived near one another because communication and transportation technologies were limited; they needed physical proximity to maintain a sense of themselves as a coherent group with a common identity. For this generation, digital technologies have made the old model nonessential.

Virtual connections are reinforced through a wide variety of conferences held at the local, state, and national levels for and by the farmers. Since 2008, the Stone Barns Center for Food and Agriculture, in Pocantico Hills, New York, has been hosting a National Young Farmers Conference in December. In 2012, some farmers who'd met at Stone Barns established the National Young Farmers Coalition, a group consisting of and working for this cohort. The NYFC works on policy initiatives, lobbies on behalf of young farmers, provides business assistance, aids farmers in accessing land, and helps young farmers create local chapters through which to build networks. It operates in concert with traditional sources of power to create new opportunities for Millennial farmers. The NYFC has made the case that farming ought to be considered a public service, much like firefighting, and that farmers should be added to the Public Service Loan Forgiveness Program.

In 2010, the Black Farmers and Gardeners Conference held the first of its annual meetings. Many Millennial back-to-the-landers are committed to equity and inclusiveness, but American agriculture—including organic and small-scale agriculture—has been a white-dominated endeavor. The policies and practices the US Department of Agriculture has long pursued, coupled with America's long legacy of interpersonal racism, have worked to exclude BIPOC farmers from landownership. In that context, claiming a land-based life can be radical and a profound expression of hope, and having opportunities for farmers of color to connect with one another is hugely important. Although BIPOC farmers comprise just 4 percent of young farmers, their numbers have grown—both in absolute terms and as a percentage—between 2007 and 2017.

Over that ten-year period, the number of young farmers rose dramatically—by 241 percent. But the number of BIPOC farmers grew even more. The number of Native Hawaiian/Pacific Islanders under the age of thirty-five rose by 495 percent, Hispanic farmers by 397 percent, and Asian-American farmers by 340 percent. Increases weren't as great for the other groups, but all except one grew more than the average. The number of African American/Black young farmers rose by 246 percent and that of Native North American young farmers rose by 244 percent. Only the percent of whites, at 238 percent, was below the average.

Like the other earlier generations of utopians described in this book, these BIPOC farmers have the sense of small-scale farming as freeing. The civil rights organizer Fannie Lou Hamer, who established Freedom Farm Cooperative in 1969, famously observed, "If you have four hundred quarts of greens and gumbo canned for the winter, no one can push you around or tell you what to do." Today, others echo that observation: For example, Dara Cooper, from the National Black Food and Justice Alliance, says people of color need land in order "to manage our own resources, feed ourselves, house our families, protect a safe space, be self-determining, and be in right relationship to earth." Many of the current "returning generation farmers" are, as Leah Penniman, of Soul Fire Farm, and Blaine Snipstal, of the Black Dirt Farming Collective, write, "community activists and popular educators who are working to dismantle this system of food apartheid and restore Black folks' rightful place of agency in the food system."[3]

Many LGBTQ+ farmers are similarly trying to dismantle the current food system and replace it with something less exclusionary. To that end, the Humble Hands Harvest, a queer worker–owned farm in Iowa, organized the Queer Farmer Network and holds an annual Queer Farmer Convergence, whose aim is "to build community among queer farmers and to reflect on and interrupt racist, capitalist, and heteropatriarchal legacies in U.S. agriculture." Likewise, Rock Steady Farm, in New York, sponsors a Northeast Queer Farmer Alliance gathering and established the POLLINATE! Program, a paid immersion experience for beginning queer and trans farmers during which they learn about farming skills and farm management.[4]

In addition to identity-based communities are others based on specific agrarian interests, such as working with draft animals or dairying. Like identity-based communities, interest-based communities rely on both in-person events and social media. Indeed, the vast majority of young farmers have embraced social media to help maintain community and to perform their utopian ideals. Andy Smith, who runs the Milkhouse Farm and Creamery, in Monmouth, Maine, with his partner, Caitlin Frame, connects with peers and gets advice from more seasoned farmers as a member of three list-servs. One, Andy says, includes "two veterinarians who have written three books on organic dairy farming. They're more than happy to respond to your random questions." Another, he says, has "some really great farmers who are willing to be resources for anybody." He was succinct about how much this matters: "We're all connected," he said simply.[5]

Digital technology also helps farmers expand their opportunities. Long before we all learned to live on Zoom, I met two aspiring farmers from France who relied heavily on digital tools to further their agricultural education. The pair were interning on farms in South and North America, working their way from Brazil to Quebec. They found the farms on the internet, interviewed for their positions via Skype, carried e-copies of farming primers and agriculture texts, and kept in touch with home by email. Farmers also use social media to advertise and expand their businesses, using websites, Facebook accounts, Instagram accounts, and the like to reach customers, potential customers, and potential employees. Early in the pandemic, when I was especially terrified, I watched more livestreams than I care to admit of the folks at Ten Apple Farm feeding their bottle-babies.

Many also use their platforms to convey their values—something earlier generations of back-to-the-landers did too. The ability of Thoreau to speak at an antislavery rally demonstrates the amplitude of his platform: He could attract a sizable audience and be taken seriously as a knowledgeable speaker on the topic. For the current generation, the platform is often digital; they're influencers who use Facebook pages or Instagram accounts or podcasts to reach audiences. In June 2020, many on Instagram posted black squares as a sign of solidarity with Black Lives Matter protesters, and plenty of farmers were among those Instagrammers.

That expression of allyship is important, but I was more inspired by the posts of Soul Fire Farm, in New York, and Maine's Four Season Farm: Soul Fire Farm posted a group photo and a link to resources for working to end racism; Clara Coleman, of Four Season Farm, began re-posting stories originally posted by Black farmers to amplify their social-media reach.

Farmers and healthy food system advocates use not only Instagram and other social-media platforms to communicate with the wider community, but also podcasts: "No-Till Market Garden Podcast" brings in guests to discuss, yes, no-till growing methods. Aaron Niederhelman's "Sourcing Matters" features conversations with food-system reformers, and livestreamed episodes of *Ask a Sista Farmer* center on the far too often marginalized voices of BIPOC, queer, poor, disabled, and immigrant people.

Thanks to digital technologies, the new farmers are adding a fresh twist to More's "utopia" pun. They live in good places—*eu-topia*—fertile land where they do good work, and in no places—*ou-topia*—the vast un-places in the digital commons where communities can coalesce. Though they may live far from one another, these Millennial back-to-the-landers are committed to community, both with one another and with their neighbors, customers, and even their land and livestock. They see farming as about more than healthy food; they see it as a means for creating more just environmental, social, and economic systems. In framing farming in ethical terms, the young farmers are taking up the utopian back-to-the-land mantle that last flourished in the 1970s.

I'm hardly the first to notice this connection. Many observers have compared these young farmers to the 1970s-era back-to-the-landers, though often to intimate that they'll probably fail in their idealist efforts. To the contrary, I—and many of the young farmers themselves—believe they have a good chance of succeeding, in no small part *because of* the seventies-era idealists. That generation put in place the pieces for an alternative food system this new generation benefits from (an accomplishment we'll explore in chapter 8). And members of this generation learned from their forebears—some by example and some quite literally. "Our generation definitely finds inspiration in the last generation of back-to-the-landers. We idealize the hippies in some ways," said Vera Fabian of Ten Mothers Farm in Cedar Grove, North Carolina. "But," she added, "we want to be more deliberate. We see that a lot of them were willy-nilly, made decisions without maybe having all the skills they needed or the resources they needed."[6]

Today's young farmers' chances of success are also greater because awareness of problems linked to industrial agriculture is more widespread, thanks in part to the environmental advocates and food proponents who have contributed to the Good Food Movement over the last several decades.

THE GOOD FOOD MOVEMENT

DEPENDING WHOM YOU ASK, the Good Food Movement is the legacy of late-twentieth-century chefs, such as Julia Child, Jacques Pépin, and James Beard, who promoted French and Italian cuisine rather than American fare, or it's a populist movement started by Farm Aid in the 1980s to help family farms survive, or it's a consumer-led uprising against processed foods and for organic and local ingredients. As a result of this amorphousness, pinning down a timeline and identifying the central players in the Good Food Movement is tricky. However, all descriptions of this movement include a few constants: a sense that interest in "good food" gained traction in the United States during the 1980s and reached wide constituencies in the 2000s; an emphasis on using local, seasonal ingredients; a rejection of processed, a-seasonal fare; and a commitment to treating food, food workers, and the environment justly. By the late 2010s, when Millennials began turning to the land to farm and create artisanal foods, the Good Food Movement had attracted a diverse array of food workers, food-justice advocates, environmentalists, and ethical consumers.

In the 1980s, life for both organic and industrial farmers was precarious. Because of federal policies and skyrocketing interest rates, some three hundred thousand farms—one in eight—in the United States went out of business. The situation was so obviously dire that in 1985, the musicians Willie Nelson, Neil Young, and John Mellencamp organized "Farm Aid: A Concert for America" to raise money to help farmers pay their mortgages. Buoyed by the concert's success, they began organizing annual concerts, at which they urged consumers to choose "good food from family farms."

Food woes weren't limited to the United States. In 1986, when a McDonald's was to be built on Rome's Spanish Steps, the journalist and gastronome Carlo Petrini staged a protest. He distributed dishes of pasta on the piazza there to emphasize the importance of traditional, cultural foods, as he put it, as a "resistance against homogenous, 'same everywhere,' fast food." Petrini didn't succeed in preventing the construction of the McDonald's, but he inspired the Slow Food Movement, which emphasizes eating local food in season and maintaining the cultural connections foods can highlight. Among Slow Food basic tenets is the belief that everyone should be able to have "good, clean, and fair food: good because it is healthy and tasty; clean because it is produced with low environmental impact and with animal welfare in mind; and fair because it respects the work of those who produce, process, and distribute it." In addition, "it is important to compensate farmworkers, fishermen, or artisanal producers in a fair way that provides them with a comfortable standard of living." (As we'll see, those tenets are central not only to Slow Foods, but also to the practices of the Millennial utopians.)[7]

Farm Aid and Slow Food paved the way, and groups with related goals came in under the large tents they erected. Environmental advocates promoted agricultural practices that were better for land, water, and animals. Food-justice activists sought more access to nutritious foods for low-income people and greater food sovereignty. Farmworker-rights and immigrant-rights activists sought better working conditions on farms and in processing plants. Even so, pricey tomatoes at the rapidly proliferating farmers' markets were easier to see than these efforts by activists, so the Good Food Movement was—and still is—often regarded as elitist.

During the 1980s, increased interest in local and seasonal foods led to a burgeoning organic-foods sector. About half of the states in the United States enacted laws defining and regulating organics. But the definition varied from state to state, which undercut consumer confidence—after all, how could something be organic in Connecticut, say, but not in Vermont? Industry leaders asked Congress to create a national organics policy. Unfortunately, the act they eventually passed, and the standards it mandated, dissatisfied almost all the stakeholders, and it took more than a decade for the USDA to put regulations into effect.

As the market for good food grew, those pursuing its various justice agendas were able to gain some ground. Items were still more expensive than industrial food, but consumers were willing to pay for foods aligned with their values—if they could be sure that farmworkers were treated fairly, for example, or that the animals were treated kindly, or that the agricultural practices were sustainable. Their willingness to pay more extended to dining out, and the idea of farm-to-table restaurants took off. Although American farm-to-table restaurants date back to Alice Waters establishing Chez Panisse in 1971, they became widespread in the 2000s. The movement was beginning to show results: For some people, the ethical and social dimensions of eating were becoming sufficiently important that they supplanted economic considerations.

In 2007, *locavore* was selected by the *New Oxford American Dictionary* as its word of the year, and in that same year three of the top ten trends identified by the National Restaurant Association were locally grown produce, organic produce, and craft/ artisan/microbrew beer. But a serious bottleneck threatened to limit how much the Good Food Movement could expand in the United States—inadequate supply. Most American farms still relied heavily on pesticides, antibiotics, and other chemical inputs. And few people were becoming farmers: In each successive Agricultural Census from 1982 to 2007, the number of "principal operators" under age thirty-five fell.

Then, Millennials began coming of age—and turning to the land. Between the 2007 Agricultural Census and the one in 2012, the number of new principal operators under thirty-five grew by an astonishing 11.3 percent. For the first time in more than thirty years, being a farmer was in vogue. Members of the largest, most well-educated generation in Western history were turning away from conventional job opportunities in favor of farming. And between 2012 and 2017, the number of farmers under thirty-five grew again—from 208,000 in 2012 to 321,000 in 2017. To get a sense of who these young farmers were and what they needed, the National Young Farmers Coalition conducted a survey. Only a fraction of those identified by the Agricultural Census as young farmers and ranchers responded, but of the 3,517 who did, 81 percent had a degree beyond high school, 75 percent didn't grow up on a farm, and 60 percent were women.[8]

I take lists of generational traits with several grains of salt, but many of the attributes associated with Millennials suggest that they're well suited to farm life: They're seen as especially conscious about social and environmental issues; are civic-oriented; skew liberal-to-progressive in their attitudes; are pragmatic idealists but also entrepreneurial; are results-oriented and good at multitasking; want to live their values; and are adventurous and self-confident.

Like Ethan and Elizabeth Siegel, other Millennial back-to-the-landers hadn't grown up intending to farm. Rather, an *aha!* moment during their teens or twenties sparked an interest in farming on a small scale and in relying on organic (or almost organic), sustainable, regenerative, permaculture, and/or Afro-Indigenous practices. Those practices are not identical, but they have quite a bit in common: All of them fit within the larger paradigm of "agroecology." Farming agroecologically involves working *with* nature rather than trying to make it conform to human will. By working *with* plants, animals, and soil, farmers enhance their well-being. In turn, the humans grow healthier and more resilient through cycles of these interactions. And these approaches benefit the rest of us, too: Agroecological methods help mitigate climate change because improving the soil enables it to sequester a lot more carbon dioxide than a depleted soil can.

THINK SMALLER, THINK SMARTER

THESE NEW FARMERS CONTRIBUTED to the increased availability of local and organic foods. Between 2008 and 2019, local food went from a five-billion-dollar segment of the food market to a twenty-billion-dollar share. Impressive as the growth of the local-food movement and of agroecological farming has been, industrial agriculture and industrially produced foods still dominate. But just as learning about GMO foods impelled Elizabeth and Ethan Seigel to start a farm, learning more about the health and environmental impacts of the industrial food system nudged Craig and Laura Martel toward starting Greener Days Farm, in Waldoboro, Maine, a town of about five thousand on the Medomak River in Lincoln County.

Their journey to the land didn't start straight out of college; they worked in the defense industry for more than a decade before feeling like they needed to change direction. Tending a garden, raising what Craig acknowledged were "illegal chickens" in their subdivision, and reading about farming set them on their path—but learning about the industrial food system solidified it.

Not long after the documentary *Food, Inc.* was released, in 2008, Craig watched it. "I actually got mad," he told me when I visited the couple's farm, then he stoked that ire by reading and watching more documentaries about the food system. "Knowledge of what our food system is like is sort of like"—he paused for a minute, searching for an analogy: "Have you seen the movie *The Matrix?* I kind of wish I took the blue pill. I'd rather not know how many things we're doing to poison ourselves on a daily basis." Mordant humor aside, he added, "One great thing about Laura and me is that when we get mad about things, we don't just sit around and stay mad. We're not victims in anything. We just go out and change it." And they did. In 2011 they purchased a woodlot and gradually transformed it into an off-grid, solar-powered farm where they raise heritage pigs. They know their farm isn't solving the whole problem, but they see it as a piece of the solution. "Maybe," Craig said, "we could change the food supply one farm at a time."[9]

During the COVID-19 pandemic, some farm and food producers did just that. Small-scale farmers and food producers were hit hard, and most of the federal Coronavirus Food Assistance Program money went to big operations. The largest 10 percent of farms received more than 60 percent of the aid money; the smallest 10 percent received a scant 0.26 percent. Without much external aid, small farms had to figure out for themselves how to weather the market disruptions.

Doing so required being quick on their feet. Joel Baranick and Ann Petricola, of Ellis Creek Farm outside Seattle, Washington, normally sold their microgreens and specialty salad mixes to wholesalers and restaurants. When restaurants closed, in March 2020, the duo cut back how much they planted and announced that they'd offer CSA shares instead. Later, as some of their wholesale clients reopened, they increased how much they were planting to meet demand. But spring and summer 2020 were full of uncertainty. As Ann explained, they had to "adapt and make changes quickly enough to be able to keep up with the changes we don't know are coming."[10]

The Maine-based cheesemaker Allison Lakin, who owns Lakin's Gorges Cheese, also sold primarily to restaurants, and in March 2020 she lost 85 percent of her client base overnight. Needing to find ways to sell more cheese directly to individuals, she created a Google Doc spreadsheet with information about her farm: what she had available, how to reach her, and so on. She added a column asking people to indicate whether they could make deliveries, then she invited other farmers and food producers to add details about what they sold and how people could buy from them safely. Within a day, she posted the information on social media.

Before the week was over, the document had morphed into an interactive map and the Cooperative Extension Service, the agricultural outreach division of the University of Maine, offered to host it on its website. Soon, traditional news outlets featured stories about it, and visits to the map and the purveyors it featured jumped. Lakin's map reflects the strong community orientation important to many farmers and artisanal-food makers. She didn't simply announce she would sell cheese at the farm; she created an opportunity for any other food maker in the state who wanted to participate.[11]

For non-farm folk, the scramble to find food during the first few months of the pandemic is something few of us will soon forget. Ordinary foods became difficult to find. Prices rose. Stores rationed. Demand at food banks rose 70 percent and newscasts showed videos of miles-long lines at food pantries around the country. At our local food bank, demand almost doubled during the spring and early summer. Americans experienced levels of food insecurity unprecedented in their memory, and this food panic heightened the anxiety people already felt about the new, deadly virus.

Goods weren't getting to the store shelves because the supply chain stuttered. In the United States, the food-supply chain is dominated by just a few companies in each sector; because they're big and powerful, they can force out or acquire smaller companies, thus getting even bigger and more powerful. In fact, when compared to the enormity of the task of getting the nation's food from field to table, the number of companies doing most of the growing, processing, and distributing is small. That might sound okay, as it suggests that the system is very efficient, but there's a flip side to efficiency: The more streamlined companies get, the more fragile they become. Enormous companies may be great at doing what they do, but they lack the robustness and resilience to change their business models if anything major goes awry. And in March 2020, everything went awry.

Again and again throughout 2020 and 2021, as assorted items disappeared from or returned to stores, I recalled a point Sara Trunzo made when I interviewed her back in 2015. For a decade, Sara worked on rural food-security issues in Waldo County, Maine, and she's passionate about the importance of local foods systems: "Being resilient through crisis is bigger than the environmental reasons to buy local food," she said. Communities need people with expertise in "growing food and managing equipment and soil, just the same way we have experts in our community who are surgeons, physical therapists, and emergency-response crews. We need that mixture of skill sets, need the tools to make food happen when and if there's a crisis like a natural disaster, a trucker strike, whatever."[12]

DREAMING THE WORLD INTO BEING

DURING THE 1970S, BACK-TO-THE-LANDERS worked to create local, resilient food systems, championing organic methods as central. By taking care of the soil and making it healthy, they could grow healthier plants. Those healthier plants could, in turn, provide more nutrients to the people who ate them. Those stronger, healthier people could further amend the soil, make it that much healthier. Those early organic advocates saw this virtuous cycle as the route to better food and greater environmental well-being. Many Millennial back-to-the-landers have adopted the 1970s-era-growers' commitment to organics-based models; 75 percent of the respondents in the National Young Farmers Coalition survey described their farming practices as sustainable and 63 percent said they farmed organically—even if they weren't certified as such.[13]

Andy Smith and Caitlin Frame, of the Milkhouse, are a great example of the sort of local food experts Sara Trunzo wants to see in her community. They've made organic principles central to their farming—and they apply a similar principle to the way they live in their human community. On a beautiful, warm May day, I headed to Monmouth, Maine, to visit them; that small, rural town is halfway between the state capital, Augusta, and the city of Lewiston. Situated on a rise on South Monmouth Road, the farm's iconic red barn was visible well before I reached the driveway. Along with the barn are their farmhouse, assorted outbuildings, and a pristine creamery where milk becomes heavy cream, yogurt, and eggnog, as well as more than two hundred acres of grass and clover pasture, where healthy cows move from field to field every other day during the snow-free months. When I got to the farm, Andy was in a near field, sporting Carhartt's, tall rubber boots, and a gimme cap that hid most of his unruly hair. In his thirties, he's one of the youngest and one of the only first-generation farmers in the area, but he's already a valued member of Maine's farming community.

After he gave me a quick tour, we sat in the couple's kitchen to talk. When he was growing up, in southern Pennsylvania, Andy didn't think highly of dairy farms; he blamed them for polluting the Chesapeake Bay through what he called their "excessive use of pesticides and synthetic fertilizers." By the end of high school, he said, he'd become aware of "a lot of problems with our food system, and its environmental impacts, and impacts on our bodies," not just those caused by the big dairies. Those realizations led him to want to farm in a healthful manner. He came to Maine to attend college and started interning on farms even before he graduated. During his first apprenticeship, his mentors introduced him to organic dairying. "Milking a cow for the first time, I really fell in love with dairy," he said, smiling.[14]

Caitlin runs the creamery side of their operation. Usually, she's there full time, but when I visited, she'd recently given birth to the couple's daughter, Willa, and was easing back into a busier farm schedule. When she and Andy began working together, they'd envisioned having a very small dairy, no more than a dozen cows, and bottling the milk and making yogurt. But twelve cows aren't enough to cover costs, so they grew the herd. Now, they care for thirty to forty milking cows, along with another fifty or so youngstock, all of which graze rotationally.

When Caitlin could pause from her chores, she came back to the house to sit and chat with us. She was effusive as she talked about the work, the farm, the chance to raise Willa and Linus, Willa's older brother, with a profound connection to place. Her appreciation for the benefits of farm life extends well beyond those for her immediate family; she's equally passionate about how it helps the land, cows, and community. Both she and Andy talked about wanting their soil to be healthy and their cows to be happy—not in a goofy way, but with the conviction that they're all partnering with one another on the farm.

To clarify what he meant, Andy rattled off some of the guidelines the USDA has for dairy farms to be certified organic; the regulations ostensibly provide a shorthand for the way cows should be treated. The cows need to be on pasture at least 120 days a year. As a rule of thumb, cows are happiest if they have access to an acre each, so the cows at the Milkhouse have more than enough room to graze fresh grass all season, which is good because another guideline is that the cows must get at least 30 percent of their "dry food intake" on pasture. Andy looked proud as he told me their cows get 80 percent on pasture. Another regulation stipulates that when the cows are in the barn, they must have "enough room to sit, walk, stretch out, and stand up without touching either any other animal or the walls of the enclosure."[15]

Those rules not only help ensure that the animals are treated humanely; they also make it possible—not easy, but possible—for small-scale dairy farmers to make a living. Dairy farmers don't set their own wholesale prices; the federal government does. So large-scale farms have a huge advantage. To sidestep that problem, smaller farms often focus on the organic dairy market, where they can get a premium for their product.

Large farms were slower to enter the organic market because getting milk certified as organic requires following a lengthy list of federal guidelines, many more than Andy and I discussed. But in December 2017, about six months after my visit to him and Caitlin, the USDA eliminated the rule about animals having to have enough room to move without touching one another. Cutting that humane regulation made it easier for dairy CAFOs—concentrated animal-feeding operations—to become certified as organic. Unlike the languid critters who loll in twos and threes in roadside fields, the dairy cows in a CAFO spend most of their lives in four-foot-by-nine-foot spaces. And these feeding operations can be huge; a large dairy CAFO has at least seven hundred cows. The USDA doesn't specify how many it can have at most; one organic dairy north of Greeley, Colorado, has more than fifteen thousand cows. Like other organic dairies, this one is supposed to comply with all of the organic requirements, but, as one dairy farmer bluntly said, "Cows move slowly and fifteen thousand would require hundreds of acres of grass per day—that's a long walk. Impossible."[16]

Even so, "organic" dairy CAFOs appear to be here to stay. In 2019, then Secretary of Agriculture Sonny Perdue said he expected factory farms to continue nudging out family dairies,

because "in America, the big get bigger and the small go out."
Perdue said this while talking with reporters after an appearance at the World Dairy Expo in Wisconsin, a state where dairy
farms were closing daily.[17]

As were many other organic farmers, Caitlin and Andy were
outraged by the undercutting of the organic standards. They
became part of a grassroots effort to establish a REAL ORGAN-
IC CERTIFICATION label for farms adhering to the core tenets of
organic agriculture. In 2019, speaking at a Real Organic Project
symposium, Caitlin urged farmers to take seriously "the full animal dignity of each creature we encounter as well as the full dignity of the land that sustains us," a sentiment that would have
been familiar to the earlier generations of back-to-the-landers.
She proposed a "bill of rights" to ensure that domestic animals
would have access to fresh pasture, air, sunlight, soil, clean food,
water, and shelter. As she pointed out, "To do anything other
than let a cow be a cow or a chicken a chicken or a pig a pig is
to create a problem for us to figure out." Caitlin concluded her
remarks humbly: "It's pretty amazing that we are grass farming,
and these beautiful, lovely cows make milk out of that grass, and
we make this really nourishing food for people out of that milk,"
she said. "It's a privilege to provide to what is now thousands
of people the food we make. Every one of those interactions is
meaningful." Meaningful, yes, and an eloquent description of
what a farm system that connects food, people, and place in a
healthy way can accomplish.[18]

It's also, at least in Andy and Caitlin's hands, a recipe for excellence: Cornucopia Institute, a nonprofit promoting family-scale
farming, rated the Milkhouse one of the top twenty organic dairies in the United States. But even that didn't protect the couple from the wave of contract cancellations that swept through
American dairies in 2018. Horizon Organic ended its multiyear
contract with the Milkhouse, giving Andy and Caitlin just six
months' notice. To weather the setback, the two innovated and

hustled, taking full advantage of the farm's resources. They added cream cheese and eggnog to the Milkhouse's offerings, and the eggnog quickly became one of its top-selling items.

Caitlin and Andy see a clear connection between local food and a vibrant community, and their wares help support their community economically and nutritionally. In turn, they eat and buy locally as much as they can, to further foster "hyper-local resilience," as Caitlin put it, which strengthens community. When customers come to the farm store, they put their money in an honor box and take their change from the till—which supports the community interpersonally by promoting trust. Andy and Caitlin also try to strengthen community by celebrating inclusiveness, often donating a portion of their proceeds to organizations working for broader social justice. And during Pride Month, the cis-gendered hetero couple fly a pride flag at the farm in allyship because, as they posted on a social-media account, "Seeing the pride flag flown on country roads is still a light for queer kids & people growing up and living in rural places."[19]

RESILIENCE AND URBAN AGRICULTURE

As were members of earlier generations of utopian experimenters, the Millennial utopians have been involved in creating vibrant urban agricultural scenes. Like their forebears, they're aware that urbanites often struggle to get access to enough nutritious food. To address that lack, they help create vacant-lot gardens and teach people about growing their own food. In this way, they enhance the availability, affordability, and quality of food in their communities.

Detroit has become well known for these efforts: The city has not only some fourteen hundred community gardens and farms, but also the nation's first "sustainable urban agrihood," a neighborhood with its own three-acre farm to help redress food insecurity established by the Michigan Urban Farming Initiative. In cities of all sizes across the nation, Millennial gardeners and farmers are transforming vacant urban lots into nutritious food hubs, turning food deserts into places where fresh foods are available.

Even before COVID-19, hunger was a serious problem in the United States—more than 10 percent of households were food insecure. Nutritious, affordable food is more difficult to obtain in neighborhoods where residents are people of color: Whereas 31 percent of white Americans have at least one supermarket in their community, only 8 percent of Black Americans do. In urban food deserts, residents often rely on prepackaged and highly processed foods sold in convenience stores or bodegas. Or they eat at fast-food restaurants, which, in contrast to grocery stores, are disproportionately located in predominantly Black neighborhoods. These fast-food restaurants are inexpensive and convenient, but most of the food they sell isn't very nutritious.[20]

Worse still, it's been designed to be habit forming. Items are calibrated to come as close as possible to what industry experts call "the bliss point," a unique combination of salt, fat, and sugar so satisfying that people crave more. Howard Moskowitz, the granddaddy of this calibration research, describes hitting the bliss points as "optimizing foods," making them the most appealing versions of themselves. Foods that hit the bliss point for at least two of those three flavors are significantly more addictive than are foods that don't, so food engineers test different ratios of fat, salt, and sugar to make items as addictive as possible. Because addictive, nutrient-poor foods are so prevalent in predominantly Black neighborhoods, diet-related ailments are responsible for ten times as many deaths for Black Americans as is physical violence.[21]

When the Winn-Dixie in the Chicora-Cherokee neighborhood of Charleston, South Carolina, closed in 2005, the area became a food desert, one of eleven in that city. More than fifteen years later, Charleston is still trying to induce a grocery store to open there. But Germaine Jenkins, who lives in this predominantly Black, lower-income neighborhood, is clear-eyed about why it's unlikely: "Grocery stores are not here because they don't make money here," she said. "They don't make margins on food. It's the nonfood items where they make their profits. If the community doesn't have the income to buy the nonfood items, then they're not going to make any money here."[22]

A few years older than the Millennials, Jenkins moved to the neighborhood in 2000, when she was twenty-five, while attending the culinary program at Johnson & Wales University. As a young mother working her way through school, she sometimes relied on SNAP or food-pantry offerings to ensure that her children got healthy foods. Though she hadn't grown up with a garden or on a farm, her commitment to nutrition led her to plant vegetables and raise chickens. Soon, she had surplus to give away. As her skills and interest continued to grow, she helped with the local grammar school's garden and a community garden. In 2014, Jenkins launched the nonprofit Fresh Future Farm on less than an acre, leasing part of a former school playground for just one dollar a year.

In 2018, I attended the annual Young Farmers Conference at Stone Barns on a media pass, which gave me access to the breakout sessions. At the time, I was part of a group trying to start a community garden to supplement our local food pantry, so when I saw that Jenkins was leading a workshop about establishing a nonprofit farm, I jumped at the chance to learn from her. Sporting a T-shirt emblazoned with the Fresh Future Farm logo, Jenkins stood in front of a standing-room-only crowd and walked us through the complicated process of getting Fresh Future Farm up and running.

First, she said, she had to clarify what, exactly, she wanted to create; she envisioned more than an expanse of herbs and vegetables. She also wanted a grocery store and a place to hold cooking demonstrations—those were essential, she said, "to show folks who are maybe used to eating most of their foods out of a box, like packaged foods, that you can incorporate some fresh produce and herbs in a way that's delicious and culturally relevant." She wanted the farm to create jobs, educate area students, and become a community hub. And she wanted it to be community driven.[23]

As the scope became more defined, Jenkins realized she possessed many, but not all, of the skills she'd need. To acquire the others, she pursued training in community revitalization, economic development, Master Gardening, place-making, and permaculture. She honed skills relevant to the nonprofit sector by being on and working with nonprofit boards. She became adept at working with city planners and on zoning issues.

Fresh Future Farm takes up about half of the school's former playground. Neat rows of produce mulched with straw abut a red-and-gray basketball court, and across from them are more plants, the chicken coop and yard, beehives, composting bins, and areas where people can linger. Less than a year after the gardens were established, Jenkins did, indeed, add a grocery store: The aquamarine building sits in one corner of the plot. In addition to produce, customers can get groceries and basic goods. Fresh Future Farm hosts farm camps for area students, trains adults in organic growing methods, brings volunteers together to work in the garden, and employs members of the community. In the first three years of operation, the farm sold more than twelve tons of groceries, below—sometimes significantly below—supermarket prices.

At the end of the workshop, the audience erupted in the kind of exuberant applause more common at a concert than at an educational workshop. Jenkins may be getting used to such reactions: She's received a prize from Southern Foodways Alliance and been named a "Southerner of the Year" by *Southern Living Magazine*, one of the "50 Most Influential People of 2018" by *Charleston Business Magazine*, and one of *Essence* magazine's "Woke 100" for 2019.

In establishing Fresh Future Farm, Jenkins wanted to invigorate her neighborhood by creating jobs, keeping money in the community, and ensuring access to culturally relevant foods. Fresh Future Farm obviously owes much to Jenkins's vision, effort, and stewardship, but it also owes much to the Chicora-Cherokee neighborhood itself. By creating solutions for and by the community, Fresh Future Farm has avoided "eco-gentrifying" the neighborhood. To the contrary, Fresh Future Farm explicitly puts residents of the neighborhood first: They get first access to produce and eggs and they can pay on a sliding scale, so the food is affordable. The store features products from BIPOC and LGBTQIA vendors, and Fresh Future Farm encourages entrepreneurship by training employees in the skills needed to pursue their own business ideas.

BACK TO OUR FOOD, BACK TO THE LAND

MY GARDEN IS CERTAINLY not a farm; however, I've also introduced many people to organically grown vegetables and legumes and have given them their first taste of fresh-off-the-cane raspberries and blackberries. When one of my brothers-in-law teased me for growing three kinds of carrots, I pulled up one of each and had him try them. He was stunned to discover that they tasted different from one another, and better than their frozen counterparts. Another brother-in-law scoffed that I'd "wasted" space on potatoes—until he ate one. No one scoffs at fresh berries; visitors of all ages eagerly pick and eat them by the handful.

Informal education was a common element of utopian experiments—especially those that were land based—and many of the back-to-the-land utopians I discuss in this book were also professional educators. Some, like Henry David Thoreau, pursued teaching careers very briefly; others, among them Bronson Alcott, Scott Nearing, Eliot Coleman, and Leah Penniman, taught professionally for a while and found ways to combine their teacherly ideals with their farming passions. That combination also animated Ben Holmes, who established the Farm School, in Athol, Massachusetts, in 1989.

Ben grew up in San Francisco and came of age at the tail end of the city's hippie heyday. In fact, he went to high school at one of the Free Schools the hippies created in Haight-Ashbury. There, he says, he absorbed "the idea of people being responsible for their own education." Some of his older friends headed back to the land to places like Drop City, in Colorado, and The Farm, in Tennessee. Ben was a bit skeptical about how well the novices would fare, as he had spent his childhood summers in Ohio on his uncles' dairy farm and knew how complex farming was. Even so, he acknowledged that the "hippie experience" had a strong influence on his vision for the Farm School, which he described to me as "very much a little slice of northern California."[24]

In creating the Farm School, he wanted to make farm experiences available to a generation who no longer had older relatives to learn from and to do so with an emphasis on personal responsibility, as in the Free Schools. Middle school and high school students can visit the working farm on class trips during the academic year and participate in extended visits during the summer; seventh- and eighth-graders can attend an on-farm full-time school. In 2001, the school added a yearlong adult program about small-scale farming and homesteading, which operated for seventeen years. Many of the adults who participated became part of the Millennial back-to-the-land movement.

Among those who spent time at the Farm School is Leah Penniman, who worked there for a year in her late teens. In 2011, with Jonah Vitale-Wolff, Leah cofounded Soul Fire Farm, an Afro-Indigenous-centered farm in Grafton, New York. Soul Fire provides opportunities for individuals with a legacy of race-based trauma to engage more safely with land and agriculture. Even as the Millennial and, now, Gen Z farmers are becoming more diverse, farming remains a predominantly white profession. And people of color interested in farming reckon with the legacies of land theft, slavery, Jim Crow, and harsh immigrant- and itinerant-labor practices as well as the challenges of farming itself. Afro-Indigenous spaces like Soul Fire Farm are a potent resource for healing hearts and repairing broken systems.

Leah's path to the land began when she worked for a summer at the Food Project, a Boston-based program that hires teenagers, teaches them to grow food, and donates the harvests to hunger-relief programs. She realized that farming could be "the intersection of the two things I loved most: social justice and the earth." That love led her to Athol and a job at the Farm School, where she cultivated more skills, and then to Barre, Massachusetts, where she co-managed Many Hands Organic Farm. As she got more involved in the organic-farming world, though, Leah says she felt less sure there was a space for her within the mostly white, mostly male movement.[25]

While at Clark University, in Worcester, Massachusetts, Leah found other opportunities to combine food-growing with social justice. She also met her husband-to-be, fellow student Jonah Vitale-Wolff. A few years after graduating, the couple, by then married with two young children, moved to Albany. To their chagrin, "it was easier to get guns and drugs than healthy food" in their neighborhood, Leah said. Determined to feed their children well, the family would walk two miles to pick up a CSA share—which cost more than their rent.[26]

Leah was teaching high school biology and environmental science when neighbors discovered that the couple had gardening skills and urged Leah and Jonah to start a farm. After saving for several years, they acquired seventy-two acres in Grafton, New York. The land is hilly, and the soil was in extremely poor condition when they bought it. Using regenerative and ancestral agricultural practices to introduce fertility, they've brought some of the land to a fecund state. While working to improve the soil, Leah and Jonah also built a post-and-beam, straw-bale house, dubbed "the Hive," in the off-hours from their full-time jobs.

Describing the work of Soul Fire Farm, Leah quotes a Ghanaian proverb: "It takes three stones to make a cooking pot stand firm." And "in honor of that we provide three services. One stone is, of course, feeding people. We use Afro-Indigenous methods to grow food for those who need it most. The second stone," she added, "is training, equipping, and inspiring the next generation of farmer-activists from the Black community. And the third is organizing to change the way resources are shared, so that Black and brown farmers will have what they need to do their work."[27]

The first stone, feeding people, is accomplished through a CSA program with a sliding-fee scale to make it broadly affordable. Team members bring shares to clients, so a lack of transportation isn't an impediment. The second stone, training a next generation of farmers and activists, occurs on farm and off. On farm, Soul Fire trains fifty-plus farmers each season, and offers workshops and community farm days. They bring urban youths of color to the farm, hoping to help them begin to reclaim a nurturing relationship to the land. Leah, and now some of the program alums, travels extensively, speaking at colleges, at community events, and at conferences. The third stone, reorganizing the way resources are made available, happens through their work advocating for land reparations, training people to see and challenge the racism within the food system, networking with allied organizations, and making sure Soul Fire Farm itself remains a safe space for farmers of color.

In 2019, in recognition of all this important work and on behalf of the entire Soul Fire Farm team, Leah received a James Beard Leadership Award.

In 2015, after serving as the founding director of the Yale Sustainable Food Project and later as the president of Slow Food USA, where he used his podium to advocate for greater social justice in the food movement, Josh Viertel established Harlem

Valley Homestead. He wanted, he said, "to connect guests with transformational experiences around living in connection with land." In 2016, Viertel hired Eric Harvey, then just twenty-seven, as Harlem Valley Homestead's first farm manager.[28]

For a semester during college, Eric had worked on an organic farm at Tenuta di Spannocchia, a historic estate in Tuscany, for a semester. There, the connections among place, food, and culture inform almost everything farmers do. Eric managed the livestock, which included a rare breed unique to the region—Cinta Senese pigs. The pigs have been granted "protected designation of origin," an appellation countries in the European Union use to acknowledge agricultural products as "of a place." Much as *jamón ibérico* is unique to Spain and sparkling wine can be called Champagne only if it comes from the Champagne region of France, Cinta Senese are unique to Tuscany.

After a year in a traditional—and dispiriting—job post college, Eric decided he wanted to farm full time. He apprenticed on Eliot Coleman's Four Season Farm to learn about raising produce, farmed in Oregon for several years, then moved to the Hudson Valley. Throughout his various farming experiences he retained a strong interest in fostering the intimate connections among place, food, and culture.

I'd known Eric casually when he was just starting out in farming, and was excited to catch up with him, in 2018, at the Stone Barns Center's Conference for Young Farmers. The conference was at full capacity, not a single room unoccupied, so Eric and I sat on the floor in a busy corridor to talk, surrounded by the happy din of people reuniting and gabbing. He had just finished his first season at Harlem Valley Homestead. The 247-acre farm was so new that it wasn't yet fully up and running; Eric mostly cared for livestock and grew vegetables. But he had plans to add orchards, berries, edible mushrooms, medicinal and culinary herbs, honeybees, and maple-syrup production. As those diverse elements get integrated, the Homestead will offer farm stays and workshops on skills for living more compatibly with the land, such as basic home gardening, food preservation, foraging, beekeeping, and seasonal cooking.

Eric described the benefits of developing intimacy with a place: "By reconnecting to agriculture and to the landscape, we become more compassionate," he said. "We gain an element of compassion, and community, and responsibility, and an ethic to do better. To eat better. To treat people better."[29]

Talking with these young farmers and food producers gives me hope that the United States might move away from the worst consequences of industrial food production. But having more local food at farmers' markets and in food deserts, although essential, isn't enough. For local food to become a viable alternative to industrial options, customers must be able to get most of their food from producers in their region. They must be able to eat locally year-round, and they must be able to afford to do so. That last is the most difficult to accomplish, even though, as I'll explain in chapter 9, the foods we buy from local producers that seem more expensive than industrially produced items often aren't.

In states with a limited growing season, preserving foods is central to ensuring year-round access to local goods. And in all states, building—or rebuilding—regional food systems is necessary. Fortunately, the production of local, preserved foods and the shoring up of regional food systems are on the rise, and Millennial utopians are involved in promoting both. Most states now permit small-scale poultry farms to butcher their own fowl on-site; some also allow on-site butchering of other livestock. Wholesale distributors for small and mid-scale producers are becoming more common. And farmers' markets now often include not only produce growers, but also meat farmers, smoked-meat purveyors, dairy farmers, cheesemakers, bakers, syrup makers, honey sellers, preserve makers, tea blenders, mushroom growers or foragers, and jam and pickle makers. Like the produce growers, these other food producers are part of the Millennial movement back to the land.

Except for the jams, jellies, and pickles found at farmers' markets (or made at home), preserved foods are seldom locally grown. But thirty-something Jenn Legnini sees preserved foods as an important component to a local diet, especially in places, like New England, that have long cold winters. Trained as a chef, Jenn became passionate about the sourcing of ingredients, and gradually changed course from working as a chef to farming. In 2013, while working on a small produce farm, she began preserving the seconds the farm couldn't sell. Creating staple preserves as well as novel items soon became her focus, and Jenn established Turtle Rock Farm in Union, a small community in Maine's Midcoast region not far from Elizabeth and Ethan Siegel's Heritage Home Farm. As Jenn switched from making preserves on the side to making them her main business, she also moved away from using seconds, and began contracting with local growers and foragers. Now, she and an all-woman crew grow heirloom produce, tend an organically managed orchard, and source from other local growers and foragers only as needed.

When I visited Turtle Rock Farm's processing kitchen, on a chilly December day, the air was suffused with the scents of rhubarb and Earl Grey tea; Jenn blends them for one of her signature fruit spreads. Turtle Rock Farm's goods are meant to last for a long time, but its processing kitchen is usually seasonal. They were making that jam in December only because rhubarb freezes well, so it could be set aside until the winter lull. Most items are processed in season: Spring, for example, is the time for pickling fiddlehead ferns. As Jenn talked about that product line, her face lit up; a local farmer who'd been making fiddlehead pickles since 1984 had contacted her to see if she'd continue making them when he was ready to retire. She felt honored that he'd given her his recipe and his foraging contacts. When the foragers bring in the fiddleheads, she said, the whole crew gets excited, as they're the first green of the year.[30]

Turtle Rock Farm makes a variety of other pickled products, such as garlic scapes, dilly beans, and cucumber pickles using a Maine heirloom, Boothby's Blond. The pickled products are complemented by whole peeled tomatoes, sauces, condiments, and fruit spreads. Some of the more unusual preserves have serendipitous histories. Because strawberries and basil arrive at the same time, usually in late June, Jenn experimented with combining them. The organic wild blueberry–cardamom spread also came about because of adjacency. Jenn had a stall at a farmers' market in between a blueberry seller and a spice merchant, and that prompted her to experiment with berry-spice combinations.

Turtle Rock Farm annually processes more than eight thousand pounds of tomatoes, nine thousand pounds of blueberries, and fourteen thousand pounds of cucumbers. Even so, the kitchen isn't in constant use, so Jenn rents it out to other growers who want to create value-added items from their own produce. When gleaners bring in the extra from fields and orchards, she converts the produce into more-shelf-stable products, turning bushels of imperfect apples into applesauce, for example. These gleanings-based items are given to food banks throughout the region. Processing kitchens like Jenn's play an important role in scaling up the viability of regional food systems. For producers, they create new revenue streams as they enable them to turn surplus into goods with a longer shelf life. And for consumers, they make more locally grown food available year-round.

Though local, organic preserved foods aren't as common as one might wish, another sector is even less well represented at farmers' markets and in regional food systems—grains. The only person I've ever seen selling flour at a farmers' market stopped after two seasons because it wasn't profitable. Truthfully, I wasn't surprised. I bought flour from him once; it was excellent, but I don't use a lot of flour when it's warm out, so I still had plenty when the market closed for the season. And local flour isn't likely to be an impulse purchase. Its scent doesn't tempt the way a sample tray of sausages might, nor does its paper sack tantalize the way golden Brandywines can.

But our diets incorporate—often depend on—whole and processed grains, so a robust regional food system should include them. Millennial farmer Ben Rooney of Wild Folk Farm began growing rice because, as he told me, "the food movement was way behind in grains" and he wanted to see what he could do to fix that. Similarly, Sam Mudge says he began growing grains in order to bring a "much-needed part of regional food production back into the Northeast." As consumers, one reason to support a local, organic-grain economy is to avoid dangerous food contaminants. When the USDA tested wheat flours, it found the neurotoxin malathion in almost half of the conventionally produced domestic samples and another neurotoxin, Chlorpyrifos ethyl, in a little more than 20 percent of those samples.[31]

THE GRAIN IN MAINE

To establish a local or regional food system, farmers and artisanal food makers must work together. So, for example, even though Turtle Rock Farm has an actual farm, Jenn Legnini isn't able to grow all of the fourteen thousand pounds of cucumbers she processes each year. Instead, she contracts with farmers, which guarantees her access to a key ingredient and guarantees the farmers an outlet for cucumbers. Similarly, cheesemaker Allison Lakin sources some of her milk because her herd of cows is too small to provide the quantity she needs. Such relationships are straightforward, but in other parts of the food system, the web of interactions can be complex. In such cases, participants have to learn what components are necessary, how they fit together, what pieces are missing, how to create the missing elements (or accommodate their absence), and how to ensure that they're working synergistically. Often, participants must also work with regional or state officials, nonprofits, philanthropists, and others. Integrating those diverse stakeholders—on top of doing the already arduous work small-scale food folk must do—can be fraught.

That complexity has not deterred the grain growers, bakers, millers, and maltsters who have embarked on an effort to (re-) create Maine's grain economy. In 2007, Amber Lambke, a Skowhegan resident who was involved with the local farmers' market, helped organize a Kneading Conference, inviting bakers, millers, and oven builders to come together, "to celebrate and inspire the idea of locally growing and milling grain for bread baked in wood-fired ovens as a way to bring the cycle of food production back to a small circle that fosters ecological and community sustainability. At the center of this idea," she noted, "is the need for partnerships between farmers, millers, oven builders, bakers, and community members." The Kneading Conference proved so successful that it's become an annual affair.[32]

Much as Ben Rooney and Sam Mudge began growing grains because they realized they were underrepresented in local food, Amber and her peers saw that local grains were missing. So, Amber said, "we set out to relearn how to grow grain in our region and inspire others to do the same." Not a farmer or a miller herself, Amber established the Maine Grain Alliance, a nonprofit hub for projects that could strengthen the region's grain economy. Somerset County, where the Maine Grain Alliance is based, was once the breadbasket of New England. In the 1830s, farms there produced enough wheat to feed a hundred thousand people. Wheat production waned mid-century, supplanted by sweet corn. By the start of the twenty-first century, the region's corn and wheat heydays were dim memories. However, the interest in local foods in the late 2000s prompted Amber Lambke and others to believe a gristmill could benefit the region. If the plan worked, it would help revitalize both the region's grain economy and Skowhegan's downtown. She called a local mill "the missing link in a local food economy that sees regional trade between farmers, bakeries, beer breweries, and raisers of livestock."[33]

When the county jail came up for sale, Amber and her business partner bought it with the intention of converting it into a mill. In 2009, when I met Amber, she and her partner were just beginning to raise money to renovate the jail; they weren't yet sure whether the mill idea would work. Amber worried they might be saddling themselves with a hard-to-sell building if it failed. But her intuition proved astute: In 2016, four years after Maine Grains opened, both it and the Maine Grain Alliance were doing so well that Amber, then director of both, had to concentrate on one and let someone else take over the other.

Tristan Noyes, a sixth-generation potato farmer, came on board as executive director of the Maine Grain Alliance at the ripe old age of thirty-two. In spring 2017, when Tristan and I connected at a local bakery in Portland to talk about the alliance's mission, he'd been its executive director for almost a year. He began explaining why the alliance was so important by contrasting the situations of grain farmers and of potato farmers. In northern Maine's Aroostook County, where he was born, more than fifty thousand acres are in potato production. There, he said, "it doesn't matter if you grow up in an agricultural family or not: If you pluck a random person out of the populace and say, 'Tell me about potatoes,' he could probably tell you twelve varieties of potatoes, how they're grown, when they're harvested, what color the flowers are, et cetera. It's built into the community itself; it's what we do." Thanks to this widely-held expertise, Tristan said, "if something goes wrong as a potato farmer, you can easily go to someone who has a lot of experience, and they can help you fix that problem." Not so with grains. That kind of knowledge, he said, "once existed in grain and no longer exists in the same way. You can't just walk out your door, go down a couple of houses, and talk to the guy who knows everything there is to know about grain growing."[34]

To fill the gaps, the alliance wants to figure out exactly what's going on with grains in Maine and locate what Tristan calls "infrastructure bottlenecks" that need to be removed. Among the bottlenecks identified early on were a lack of drying methods, sorting methods, and storage units that would work well in Maine. Confusion must have been written on my face because Tristan immediately elaborated, explaining that the various tools for those tasks are made with specific environments in mind. Equipment designed for the Midwest may not work well in Maine, he said, because Maine is wetter and its temperatures fluctuate more. Instead of adopting methods from the heartland, Mainers have been going to Denmark, where the environment and weather are more like New England's, and where the regional grain scene is considerably farther along.

I wanted to know why a potato farmer was interested in heading a grain organization, and Tristan said it was precisely because he was a potato farmer. Like most other farmers in the region, Tristan and his brother, Jon, rotate their potatoes with grains to control for pests and replenish the soil. But grains bring in barely 5 percent as much money per acre as potatoes do, so lots of farmers in Aroostook County plow them under or sell them to commodity markets. A few years earlier, Tristan and Jon had decided to convert some family land from conventional to organic production, which meant they'd have to use organic grains in their rotation. At the time, all they knew about organic grain was that it cost more—which gave them an idea. If farmers grew potatoes organically, then their potatoes would sell for a premium and they could also sell their rotation crops for more.

Potato farmers routinely use spring and winter wheats, buckwheat, oats, or barley in their rotations. Keep barley in the back of your mind; I'll come back to it in a little while. But first, the wheats. In principle, the farmers can plant common wheat, grow it organically, and charge a little more for it than the industrially grown version sells for. But if they grow older or unusual species, they accomplish more. The less-common grains fetch a higher price than they'd get for common wheat, and growing them helps keep the species commercially viable.

Believing it would grow well in northern Maine, Will Bonsall, a seed saver who got his start during the 1970s-era back-to-the-land movement, gave the alliance fourteen pounds of Sirvinta, a rare heritage winter wheat from Estonia that's easier to digest than are newer strains of wheat. In fact, some people with gluten sensitivities can eat it without any problems. Through its Seed Restoration Program, the alliance partnered with farmers and home gardeners to grow out the Sirvinta and saved the seeds from the hardiest, best-yielding plants. By 2018, the alliance had enough Sirvinta to begin offering it commercially: "We have the largest supply in the western hemisphere," Tristan told me proudly.

Whereas grain grasses are the most common rotation plants for potato farmers, beans are particularly good rotation crops for grain growers. So in addition to selling flour milled from the grains its growers produce, Maine Grains has begun selling the beans the grain farmers plant in alternate years, including organically produced Jacob's Cattle, Marfax, pinto, black, and yellow beans. In 2021, two Maine Grain products, Pearled Black Barley and Marfax Crop Rotation Dry Beans, each earned a Good Food Award, a national accolade for delicious, sustainably produced foods. That Maine Grain won one for a rotation crop as well as for one of its grains makes clear how well it has done to help farmers integrate higher-value rotational crops into their farm plans. Tristan's good idea is bearing fruit—or legumes, as it were.

Even so, ensuring that a regional grain system is resilient will depend on having more mills processing local grains, more eaters enjoying them, more brewers using them, and more bakers cooking with them. During spring 2020, that last came true. As pandemic baking took off, sales rose significantly more for regional flour companies than for big labels like Gold Medal. Maine Grains had a *4,000 percent increase* in online sales. Flour took on heightened emotional resonance during the pandemic; people who hadn't cared where their flour came from were investing more of themselves in baking and wanting more out of it. Impressed by the uptick in sales that Maine Grains and other small flour mills experienced, Tim Wu wrote in the *New York Times* that regional mills could be a model for other parts of the economy: They offer distinctive products, are often employee owned, and employ more people relative to the amount they produce.

Cereal grains have appeal beyond the plate. In 2013, another big-picture thinker, Joel Alex, founded Blue Ox Malthouse to support Maine's grain economy and to supply craft brewers with a local malt. Blue Ox opened a commercially scaled operation two years later in a small industrial park in Lisbon Falls—a town best known as the inspiration for Stephen King's *Castle Rock*. When I arrived to tour it and learn about malting, Joel was outside by the loading dock of a plain, tan building. Wiry, clad in a plaid shirt, with an intense gaze his glasses did nothing to obscure, Joel looked more like a student from my art-school days than any of the young farmers I'd been getting to know of late. Inside, I beheld what seemed a cavernous space—seventy-five hundred square feet with a ceiling roughly three stories up.

When I marveled at the size, Joel said that as malthouses go, Blue Ox is actually very small. As we walked, he gave me a primer on malting, starting with the fact that Blue Ox "floor-malts" grain, the method used for thousands of years but which is now rare. At Blue Ox, the process begins with a batch of grain weighing between five and six tons. Mostly, it uses barley—one of the grains the potato farmers use as a rotational crop. The maltsters steep the barley in water until it reaches the right moisture level, which usually takes a couple of days. Then they lay the grain on the malt floor, a section of the warehouse that's been separated from the rest and is climate controlled.

The grain stretches almost wall to wall and forms a plush carpet about four inches deep. It needs to germinate just enough to release certain enzymes but not enough to start sprouting, a process that takes about four days. Workers use malting rakes to turn it over three or four times a day; that way, the tiny rootlets on the germinating grain don't twist together and mat and the grains all stay at the same temperature.

I couldn't get over how beautiful the grain looked. The swells and troughs of the recently raked grain were in counterpoint to corrugated metal walls. It reminded me of the undulating waves of sand in a Zen garden. Joel agreed, though he did point out that raking wet grain is much more difficult and much less contemplative than raking dry sand.

Once the grain reaches the proper germination state, it's shoveled onto a conveyor belt and fed into a kiln to dry slowly. Kilning lacks the visual appeal of the malt floor, but it's during this step that much of the malting magic happens. Changing the duration and/or intensity of the heat applied during kilning modifies the flavor and color, so maltsters tweak them to add complexity. After the grain is kilned, Blue Ox smokes some malts to introduce still more flavor.

Floor malting was the default process until the nineteenth century, when malting was mechanized. By the end of the twentieth century, every aspect of the malting process had been automated; the machines can even clean themselves. Now, large malthouses can process more than a billion pounds of grain a year—a thousand times more than Blue Ox can. But as with so many other agricultural and food products, the trade-off is variety and flavor. Fewer than ten companies produce more than 95 percent of the world's malt, and they offer uniform products designed to meet the needs of domestic beer producers.

Blue Ox, in contrast, can experiment with its recipes, adjusting the kilning, roasting, and smoking steps, making different combinations to develop unique malts with varied flavor profiles. This capacity to make unique products has contributed to the craft malthouse renaissance. When Blue Ox was founded, there were just five operating craft malthouses nationwide. By January 2020, that number had ballooned to one hundred. The pandemic dealt craft maltsters a blow, as bars and tasting rooms were closed or operated at reduced capacity for much of 2020; Blue Ox is among the fifty or so survivors. Like Blue Ox, most of the other craft malthouses that prevailed are in states either with a strong local food movement, like North Carolina, New York, and Pennsylvania, or with a high number of craft brewers, like Michigan and Montana, or with both, like Maine.

Craft brewers are Blue Ox's target market, but Joel says his goals for the company go well beyond providing a great product. Joel is someone who sees how and where to modify a system to make it more robust and resilient. In college, he majored in environmental studies and international studies, "but in retrospect," he said, "I was studying development." The more he learned, the more he realized that traditional development models can't simultaneously honor the uniqueness of a place and its culture and be sustainable. Genuinely sustainable development requires considering ecological, social, and economic sustainability. "You kind of need all three," Joel said, "and if you look at our society, we're none of those. We're not ecologically sustainable, socially sustainable, or economically sustainable."[35]

True, as a society, we're not. But Joel knew it was possible, at least at the local or regional level, and he wanted to find ways to support the region's agricultural economy by scaling up its sustainability. Creating opportunities for farmers to add value to their products—much as the Maine Grain Alliance helped farmers add and market their rotation crops—would enhance farm resilience. To learn how to do this, Joel intended to go to graduate school for an MBA in sustainable development. But a conversation in 2012 with a local craft brewer derailed that plan. At the time, Maine grain was shipped to Canada to be processed, reimported by a distributor from New York, and then shipped back to Maine. As the pair talked, Joel told me, "all kinds of lightbulbs went off." A local malthouse could provide an outlet for organic grain farmers, encourage Aroostook County farmers to grow malting barley as their potato rotation crop, supply the booming number of craft breweries with the local products they sought, add infrastructure to enhance farmers' access to what he calls "mid-sized premium markets," *and* employ people. The pieces fit together; he believed the plan would work. Joel skipped graduate school and built Blue Ox.

Not that it was easy. At age twenty-seven, with virtually no credit history, no business experience, no experience making malt, and no way to demonstrate that a sizable market existed for a product that hadn't been offered in the United States in a century, he was "not a super-financeable person," as he wryly put it. He was, in fact, the opposite of super-financeable. He wasn't eligible for bank loans and didn't have personal funds or friends with deep pockets. So Joel turned to organizations in Maine that promote impact investing. These public and private groups work with aspiring entrepreneurs who have ideas for businesses that provide a social good as well as a financial return. He was diligent about finding mentors and financial resources; even so, building a malthouse entails so many upfront costs that Joel lived out of his car for eighteen months while getting Blue Ox up and running.

Although still very much a startup, Blue Ox (and Joel) is on firmer financial footing now. The company has nine employees, buys grain from ten Maine farms, and sells malt to almost a hundred New England brewers, distillers, and homebrew suppliers. Creating good jobs with benefits, adding infrastructure to the grain economy, and offering a high-quality product made in an environmentally kind way, Blue Ox strives to embody the three criteria for sustainability Joel had listed. And it's helping to spark other virtuous cycles in the grain economy.

As we walked by the pallets of finished malt, I saw "Allagash" scrawled with a Sharpie on quite a few. Joel confirmed they were destined for the Allagash Brewing Company. Founded in 1995,

Allagash is one of Maine's old guard of craft brewers and one of the biggest, producing nine hundred thousand gallons of beer each year. Over the past decade, the handful of first-generation craft brewers have been joined by roughly a hundred second-gen brewers. *Beer Connoisseur* magazine credits Allagash and its founder, Rob Tod, with inspiring some of that influx: Allagash helped "set the stage for a bundle of other breweries pouring delicious funky and sour beers."[36]

Allagash's commitment to funky beers is coupled with one to community. Over the years, it has given back in many ways. Like many other craft brewers, it gives its spent grain to livestock farmers: "It's just always been a very symbiotic relationship between a brewer and a farmer," explained Jason Perkins, a brewmaster for Allagash. Both brewery and farmer benefit; the brewery doesn't have to pay to dispose of spent grain, and the farmer saves on grain costs while adding variety to the animals' diets. Perhaps the most novel way Allagash is giving back is through its promise to source a million pounds of Maine-grown grain annually. Knowing that a guaranteed market exists will make it easier for entrepreneurs like Joel Alex to take chances and for Tristan Noyes to convince additional farmers to grow organic grains.[37]

Working in concert, farmers, millers, brewers, and maltsters have reestablished a regional grain system, with plenty of help along the way from bakers, seed savers, and entrepreneurial risk-takers. Some of the participants—such as Amber Lambke and Rob Tod—aren't Millennials, but many, many others, including Jon and Tristan Noyes, Sam Mudge, Ben Rooney, Joel Alex, and dozens of craft beer makers, are. Their energy and ideas are fleshing out the model Amber envisioned of a regional and integrated grain economy that benefits farmers and communities. The system is a work in progress, and problems will continue to crop up as it grows, but by facing them together—being in community—the participants have a better chance of tackling new challenges and Maine has a regional grain system that creates excellent products, is kind to the environment, and keeps money circulating locally.

This integrated grain economy can serve as a model for others. Whereas craft brewers eagerly embraced local grain and hops options, distilleries have been slow to do so. Moreover, as Shanna Farrell observes in *A Good Drink: In Pursuit of Sustainable Spirits*, their agricultural practices are often unsustainable. But new approaches are beginning to take hold. In the Hudson Valley, farmers, maltsters, and bakers have been working together since 2015 to increase the quality and availability of local grain. And in 2019, the spirits maker Brown-Forman began partnering with farmers, University of Kentucky crop scientists, the American Farmland Trust, and others to bring rye back to Kentucky farm fields, initially as a cover crop, but with the hope of having a local rye that Woodford Reserve, one of its distillers, can use in bourbons.

Bourbon's been having a moment—well, a decade—and Brown-Forman believes local rye would add even more to its appeal (and reduce shipping costs). If the University of Kentucky's crop scientists can find or develop a rye tolerant of the region's warming temperatures, then this collaboration among grain farmers, distillers, livestock farmers, and Kentucky bourbon aficionados will further strengthen the region's grain system.

MOVING AHEAD

MILLENNIAL BACK-TO-THE-LANDERS—THE FARMERS, artisanal food makers, and their allies—are adding a rich array of local products to our food system. At the same time, they're making strides toward improving the health of the land where they work and the healthfulness of the food available to their communities, as well as educating folks who come into their orbits about the importance of such efforts. Even so, it remains an open question whether this generation can resuscitate local food systems and food cultures. The National Young Farmers Coalition maintains this generation *must* transform America's agricultural system, but accomplishing that is even more challenging than is farming successfully. The federal government will need to take a decisive role, one that involves reducing its current support for large-scale industrial agriculture and increasing its support for small-scale agriculture; entrenched corporations will need to cede power or be pushed aside; and consumers will need to pay more for food at the cash register. Believing one generation can overcome all of those hurdles does sound utopian.

Still, it isn't all tilting at windmills. Utopians remind us that alternatives exist. As the foods Americans grew and ate became more monolithic, the agriculture methods more uniform, and the corporate ownership more consolidated, utopian challenges of the sort chronicled here became ever more important—demonstrating ways to reinvigorate the soil, the community, the food system, and the economy. As they nestle into their communities, young farmers both revitalize and unsettle them. In that, too, they're like their predecessors, whose presence sometimes discomfited, sometimes inspired, and always revealed the existence of fresh options.

2

HENRY DAVID THOREAU'S
SEARCH FOR FREEDOM

M OST UTOPIANS WHO GO back to the land don't become famous for it, but a few have become cultural icons— none more so than Henry David Thoreau. In 1845, he settled into a hand-built cabin near the edge of Walden Pond, in Concord, Massachusetts, where he lived for two years and two months. But he was far from the only one in his day going back to the land in order to determine how best to live. In 1840, Ralph Waldo Emerson drolly noted in a letter to his friend Thomas Carlyle, "We are all a little wild here with numberless projects of social reform. . . . One man renounces the use of animal food; and another of coin; and another of domestic hired service; and another of the State; and on the whole we have a commendable share of reason and hope." During that decade, some eighty utopian experiments were under way and tens of thousands of people participated. Until the 1970s back-to-the-land movement, it was regarded as the golden age of utopian experiments.[1]

Several of Thoreau's close friends also participated in utopian back-to-the-land experiments, attempting to create heavens on earth where participants could manifest their natural divinity. Likewise, other utopians, such as the Shakers, the Oneidans, the

Icarians, and the Fourierists, created small communities and hit restart. They believed that if people could design a culture and society *de novo*, following a different philosophy than that shaping the mainstream, they could avoid its problems and attain human perfection. What the founders of the utopian movements saw as the most serious social woes helped frame the particulars of their various philosophies: Some considered gender inequality to be the biggest impediment to flourishing; others saw racial inequality as the main issue. Some regarded capitalism as the central problem, or rising industrialism, or corrupt institutions. All had a sense that America's promise of freedom was on the wane, leaving people unable to realize their full potential.

Thoreau had much in common with both his peers and subsequent back-to-the-landers, most importantly the desire to define, and dwell in, freedom. But he was also the odd man out, as he lived alone during his utopian experiment, whereas his utopian peers pursued communal quests. All the same, his effort has had the most lasting impact thanks to his book *Walden; or Life in the Woods*, which came out in 1854. At the time, it received good reviews, but it wasn't an immediate best-seller by any means. One early reviewer accurately predicted that "the influence which Mr. Thoreau exerts will not at once spread over a large surface, but it will reach far out into the tide of time, and it will make up in depth for what it wants in extent." Sure enough, since 1862, *Walden* has been continuously in print, and its significance only grows: Currently, some seventy-five different editions of it appear in English, and in 2004, during the sesquicentennial celebration of *Walden*'s publication, one scholar pointed out that "*Walden* is arguably the most translated work of American literature in terms of the number of editions in foreign languages and the number of languages in which it has been translated." It was a cherished essential text to those who went back to the land around 1900, during the Great Depression, and in the 1970s.[2]

Like all successful utopian texts, *Walden* answers the question of how someone who rejects prevailing cultural values might live a good life. Thoreau's solution was grounded in a place, connected to a community of living beings, infused with intellectual rigor, and enabling of freedom. As American culture began moving more toward a market economy, in the early 1840s, being free was increasingly in tension with making a living, so Thoreau also included a good bit of economic critique, along with philosophical rhapsodizing, dietary ethics, playful observations, and the paeans to nature that have made him a hero to successive generations. What *Walden* doesn't provide is much practical guidance about the day-to-day matters of simple living.

Walden became acclaimed as the prototypical back-to-the-land book; in its wake, a new genre emerged—accounts of an individual or a group who learn to live simply on the outskirts of society. Scott and Helen Nearing, who went back to the land in the 1930s and became known as the grandparents of the seventies-era back-to-the-landers, quoted Thoreau quite frequently in their books. When I visited the couple's former home, I found an early edition of *Walden* among their books, a beech leaf still tucked between its front pages. Helen Nearing so admired Thoreau that she asked for her papers to be preserved at the archive where his are. And Stewart Brand, who created *The Last Whole Earth Catalog* to help hippie do-it-yourselfers find the tools they needed, described *Walden* as "the prime document of America's third revolution, now under way." Thoreau's influence extends beyond those who dream of homesteading: He is credited with inaugurating America's environmental movement.[3]

The first intimations of the philosophy that permeates *Walden* came not from coursework or from conversations Thoreau had during his undergraduate years at Harvard College, but from the half-semester he spent living with Orestes Brownson, a philosophically and socially radical Unitarian minister. During his junior year, Thoreau took a semester off to earn enough money to finish school. Brownson hired him to teach grammar school in Canton, Massachusetts, and this detour rocked Thoreau's world. For six weeks, he lived with the minister and his family. Brownson was a brilliant thinker, eager to talk, argue, prod, and probe, and Thoreau reveled in it. He later told Brownson their time together was "an era in my life—the morning of a new *Lebenstag*," the first day of the rest of his life.[4]

I don't know exactly what the two talked about, as Thoreau had not yet begun keeping his detailed journal, but I can guess. Brownson was then working on the manuscript of *New Views of Society, Christianity, and the Church*, which was published the following year. Twice in its preface Brownson reminds readers that the views he expresses are original—which is important to keep in mind, as his ideas would, in time, become very familiar. Brownson's radical ideas, along with those in Ralph Waldo Emerson's essay "Nature," which came out that same year, are at the core of the philosophy they and others were collectively articulating: transcendentalism.

In *New Views*, Brownson maintains that in ancient cultures, religion had become "a function of the state"; when that happens, people have a hard time claiming individual authority. In contrast, he said, the revolutionaries who founded the United States had wrested their individual authority from England and the church. With one taste of freedom, he noted, came the desire for another, for "freedom of thought and conscience." Brownson didn't believe Americans had attained those additional freedoms. Instead, he considered the people of his era too materialistic; he lamented that "men labor six days for this world and at most but one for the world to come." To him, this undermined the promise of America as a "new paradise [that] was imaged forth for man."[5]

But Brownson felt a repudiation of materialism was under way; in some circles, "money-getting, desire for worldly wealth and renown, are spoken of with contempt." In the place once occupied by material desires, he sensed a longing for something more "soul-kindling." People yearned for a greater awareness of the divine and a deeper sense of their own nature as itself divine. The new doctrine he was framing could, he wrote, slake those yearnings, for it taught "that spirit is real and holy, that matter is real and holy, that God is holy, and that man is holy." People who really believed this would treat one another with greater respect and kindness because they would recognize each person's divinity. Likewise, they would see earth and animals as sacred, to be used only in a holy manner. Brownson recognized such a huge transformation of consciousness wouldn't happen quickly, but he believed it would happen.[6]

Thoreau was captivated by Brownson's ideas; he read *New Views* as soon as it came out. By graduation, he'd adopted many of Brownson's conclusions, as we can see in the commencement address he gave at Harvard College in August 1837. Assigned to speak about "the commercial spirit of modern times, considered in its influence on the moral character of the nation," Thoreau managed to turn the topic on its side. When the rains finally abated that steamy morning, he recited his short speech, telling his classmates, teachers, and other attendees that "the characteristic of our era is perfect freedom,—freedom of thought and action."

Despite the boosterish sound of this opening, Thoreau didn't celebrate the man of commerce. Instead, he excoriated the commercial spirit for impelling "a blind and unmanly love of wealth" that infuses all actions with selfishness. He reassured his audience that such selfishness could be overcome, that those who were true to their natures, who freed themselves from the strictures of striving, could enjoy an earthly paradise. And in his formula for attaining this "nether paradise," we see not just a recap of Brownson's general ideas, but also his specific point regarding toil and the Sabbath:

> This curious world which we inhabit is more wonderful than it is convenient; more beautiful than it

is useful; it is more to be admired and enjoyed than used. The order of things should be somewhat reversed; the seventh should be man's day of toil, wherein to earn his living by the sweat of his brow; and the other six his Sabbath of the affections and the soul, in which to range this widespread garden, and drink in the soft influences and sublime revelations of Nature . . . [and] pass the days of his sojourn in this his nether paradise.[7]

After graduation, Thoreau returned to live with his family in Concord, a crossroads for traders thirty-five miles west of Boston. Thoreau was soon immersed in two of the most potent progressive movements of the era. Thanks to his mother's staunch abolitionism, his family's home, which was also a boardinghouse, had become the unofficial headquarters for Concord abolitionists. And Ralph Waldo Emerson's home, not even a mile away, was the base of an emerging intellectual movement. Emerson and several other Unitarian ministers had organized a study group to reconsider certain aspects of Unitarianism, much as Orestes Brownson was doing, and to foster more philosophical thinking in America. Emerson urged his peers to invite Amos Bronson Alcott, even though he wasn't a minister, because Emerson considered him "a God-made priest." The others acquiesced, and when the group next met, they invited not only Alcott, but also Orestes Brownson and several others.[8]

The following winter, Thoreau was invited to join their group, which was articulating a philosophy emphasizing the inherent divinity of all beings, the connectedness of humans to the rest of nature, and a faith that such knowledge might reform the world. The other participants became some of Thoreau's closest friends, and the views they collectively cultivated would shape Thoreau's philosophy and actions, especially during his time at Walden Pond. In fact, two other members of the group,

Bronson Alcott and George Ripley, created utopian communities steeped in transcendental ideas.

TRANSCENDENTALISM

DURING THE 1990s, I taught American literature at a liberal arts college in Pennsylvania. Every year, at least one of my courses incorporated a hefty measure of transcendental literature, usually essays by Emerson and Thoreau. And every year, one or two students were beguiled by their ideas. My very first semester, one student was so fired up by reading sections of *Walden* that he became a fruitarian and began living in a tree. (Strictly speaking, Thoreau recommended neither. He did write "There is some of the same fitness in a man's building his own house that there is in a bird's building its own nest," but I don't think that qualifies as a call to treetop living.) Over the rest of the decade, no one else fell so in love with transcendentalism. To the contrary, many students dismissed it as wifty.[9]

Now, with plenty of hindsight, I get it.

Transcendentalism is about freedom and lauds the individual as his or her own best teacher, able to perceive by intuition what is true. It calls for following the individual conscience rather than rules and laws, as institutions tend to corrupt one's basic goodness; and it praises thinking outside the box and reveling in one's nonconformity. Alongside this emphasis on individualism and independence is an understanding of the world as profoundly interconnected. Humans are part of a oneness with all others, part of the "web of relations to all beings . . . which makes it impossible for [one] to act without affecting many more than one." Emerson said this oneness often seemed lost, and transcendentalism offered a way to reclaim it.[10]

Getting graded on how well one speaks or writes about reveling in nonconformity, or trusting one's intuition, or embracing one's inherent divinity seemed ironic, at best, to my students. As did discussing the importance of learning from nature and observing its rhythms and principles while sitting on plastic chairs in an over-lit, overheated, mostly monochrome room. Even if we joked away the ironies, other transcendental beliefs were tough for those students to take seriously. Emerson urged readers to spend their time contemplating beauty and assured them the world was not something out there, but rather something they created. For the science majors taking the class to fulfill a humanities requirement, that seemed like nonsense. And Thoreau? He rejected almost everything most of them had entered college wanting to achieve—a professional career, social standing, financial comfort. Coming of age during the economic boom years of the nineties, these students were far different, temperamentally, from either their Baby Boomer parents or the transcendentalists.

The Boomer generation included hippies who kept their dogeared copies of *Walden* tucked alongside their *Whole Earth Catalog*s and issues of *Mother Earth News*. For them, Brownson's assurance "that spirit is real and holy, that matter is real and holy, that God is holy, and that man is holy" would have induced not the eyerolls it got from their kids, but, rather, a sense of déjà vu. His words sounded like a tamer version of the Beat poet Allen Ginsberg's ecstatic avowal that "The world is holy! The soul is holy! The skin is holy! The nose is holy! The tongue and cock and hand and asshole holy! Everything is holy! everybody's holy! everywhere is holy!"[11]

Even though I wanted the students to love Thoreau, I had to admit they were doing what he explicitly urged—trusting their own intuitions and spirits and ignoring those of their elders. In *Walden*, Thoreau wrote, "Practically, the old have no very important advice to give the young, their own experience has been so partial, and their lives have been such miserable failures." When I was on my students' side of thirty, I quite liked Thoreau's snippy aphorisms. I even had a poster about marching to the beat of a different drummer in my college dorm room freshman year. So it came as a bit of a surprise to discover that Thoreau's own behavior often ran counter to the nonconformist image he cultivated.[12]

In addition to being so inspired by Brownson that he quickly adopted many of the older man's ideas, Thoreau copied Ralph Waldo Emerson and Bronson Alcott. He aped Emerson's language patterns and mannerisms so deftly that one of his friends couldn't tell them apart: "In the tones and inflections of his voice, in his modes of expression; even in the hesitations and pauses of his speech, he had become the counterpart of Mr. Emerson," the friend recalled, adding that he was "so much struck with the change" that he closed his eyes and found he was "unable to determine with certainty which was speaking."[13]

Although Thoreau imitated Emerson's voice and mannerisms, he never achieved the physical grace his mentor displayed. To the contrary, poor Thoreau was—at least in Nathaniel Hawthorne's opinion—"as ugly as sin; long-nosed, queer-mouthed, and with uncouth and somewhat rustic manners." Hawthorne didn't mean to be unkind, apparently, for he went on to say "his ugliness is of an honest and agreeable fashion, and becomes him much better than beauty."[14]

Thoreau mirrored Alcott in both thought and deed. In January 1843, Bronson Alcott was briefly jailed for refusing to pay his poll tax in protest of the planned annexation of Texas. Admiring this exercise of conviction, Thoreau decided to stop paying his own poll tax. Nothing happened until 1846, when he, too, was jailed briefly for refusing. Thoreau's defiance is far, far better known than Alcott's, recounted eloquently in the essay "Civil Disobedience." Similarly, when Thoreau lived at Walden Pond, his beliefs and day-to-day practices echoed Bronson Alcott's: Both sought to live self-sufficiently, and thought that by doing so they could reclaim their best natures. Both thought being divorced from the capitalist, consumerist culture was essential to that effort. Both sought to avoid exploiting others. Both believed they should avoid stimulants and alcohol, follow a largely vegetarian diet, eat whole-grain breads, limit their use of imported goods, and grow much of their own food. Their food choices, like those of utopian back-to-the-landers in later generations, reveal aspects of the agricultural practices of their era they found troubling.

In pointing out how much Thoreau borrowed from his friends, I'm not trying to deride him or dismiss his contributions. Rather, I want to point out that although he was a loner, he grew up in a close-knit and devoted family whose values and ideals he held, then became part of an intellectual circle whose values and ideals he also held. Before, during, and after his time living alone at Walden Pond, he had the good fortune to be part of not one, but two mutually constitutive communities. Young Thoreau's copying of his older mentors suggests how much he admired the friends with whom he was co-creating the heady froth of transcendental thought.

MOVING TO WALDEN

WHEN THOREAU WAS TWENTY-SEVEN, his friend Isaac Hecker invited him to travel around Europe on a walking tour, which was Thoreau's favorite way to travel. Even so, he couldn't bring himself to go. Instead, inspired by the several weeks he'd spent living in a "shanty" beside Flint's Pond with his college roommate, Thoreau wanted to move to a shanty of his own and write a book about a two-week journey he and his best friend—his beloved older brother, John—had taken. The book was to be a memorial to John, who had died of tetanus, in Henry's arms, in January 1842.

Thoreau probably also yearned for some peace and quiet. The boardinghouse his mother ran was genial but often loud. And throughout Concord, he was taunted by people muttering "Burnt Woods" when he passed by. The year before, he and a friend had accidentally started a fire that destroyed almost one hundred and fifty acres of Concord Woods, and his neighbors didn't let him forget it. So in March 1845, when William Ellery Channing wrote to him saying "I see nothing for you in this earth but that field which I once christened 'Briars'; go out upon that, build yourself a hut, & there begin the grand process of devouring yourself alive," it must have seemed a perfect solution. After all, Thoreau's good friend Emerson owned Briars, a plot of about fourteen acres. Emerson agreed to let Thoreau squat there, on the edge of Walden Pond, in exchange for improving the property.[15]

The pond had long been one of Thoreau's favorite places. He loved boating in general, but especially on the kettle pond. Created by retreating glaciers more than ten thousand years earlier, Walden Pond water was so clear and pure in Thoreau's day that ice was harvested from it both for local use and for markets around the world. And though its shores were scantly populated, the pond was not remote. Concord was just a saunter away; the Thoreau family lived barely two miles from Briars. The Fitchburg Railroad, which had opened a station in Concord a year earlier, was adding more track not far from Thoreau's spot. Glad that the railroad enabled easy access to the library at Harvard College, Thoreau was less pleased to hear the train's roar drowning out birdsong and other natural sounds.

Within weeks of securing Emerson's permission to use the land, Thoreau borrowed an ax from Bronson Alcott so he could "cut down some tall arrowy white pines . . . for timber" and get started building. Over the next several months, he borrowed more tools, cleared more land, dug a cellar hole six feet by six feet and seven feet deep, and framed his cabin. He purchased a shanty from James Collins, an Irish immigrant working on the railroad, and dismantled it for wood and nails. For the chimney, he used recycled bricks and stones he carted up from the pond. In May, he asked a few friends to help with the house-raising—"rather to improve so good an occasion for neighborliness than from any necessity," or so he said—and soon had a watertight ten-foot-by fifteen-foot home. Come fall, when the weather grew cool, he plastered the walls, shingled the outside, and finished the chimney.[16]

As soon as the ground could be worked, Thoreau began his garden, hiring a driver and a team of oxen to help him plow two and a half acres for vegetables. They pulled so many stumps that Thoreau had almost enough firewood for the two winters he spent there. Once the fields were prepared, he planted peas, potatoes, turnips, sweet corn, yellow corn, and—wait for it—seven miles of bush beans. If he'd had a good yield, he may well have ended up with more than three thousand pounds of white beans, an absurd quantity. Even losing much of his harvest to a woodchuck, he ended up with 720 pounds. Given my own beany passions, I couldn't help noticing that this excellent, practiced gardener, the one who urged others to "simplify, simplify, simplify," set himself up for a lot of unnecessary bean-tending. More on that later.

Thoreau moved into his cabin at Walden Pond on Independence Day 1845 and stayed there for two years, two months, and two days—except that night in jail and about two weeks in Maine. In "The Ponds" section of *Walden*, he draws a parallel between Walden Pond and the Garden of Eden; like so many others of his era, he regarded his experiment as an opportunity to create an earthly paradise, the "nether paradise" of his commencement address. Specifically, he wanted to simplify his life so he could find the precise circumstances for flourishing. Early in *Walden*, he explains that he sought "to live deliberately, to front only the essential facts of life, and see if I could not learn what it had to teach, and not, when I came to die, discover that I had not lived. I did not wish to live what was not life, living is so dear." Fronting those essentials meant spending more time in nature looking, thinking, talking, walking, and writing, and less time engaged in paid labor.[17]

Some of his neighbors considered Thoreau lazy for not wanting to work all the time; one described him as a "good-for-nothing, selfish, crab-like sort of chap who tries to shirk his duties." But his distaste for paid labor wasn't a character flaw. Rather, he believed people were enslaved to their property and their desires. By figuring out what he could live without, he could spend less time laboring and more time being free. He was confident he could earn enough in six weeks to live for a year and intended to show others it could be done. Persuading his contemporaries was going to be a hard sell: Most farmers worked at least ten hours a day, six or seven days a week, and people in manufacturing worked almost seventy hours a week.[18]

His prospects for cultivating this lifestyle—if not of convincing others to try it—were aided by Emerson lending him land and Alcott an ax, relieving him of one large and one small initial expense. Many of his minimal furnishings were also borrowed, as he believed property encumbers. He pitied his neighbors who inherited farms; too many were "serfs of the soil," obliged to work maintaining property they had not sought. Many of those family farms had been successful for decades but were suddenly unable to run as they had.

When Thoreau was coming of age, farming culture in Concord, as in much of the rest of New England and in the Midwest, was changing dramatically. The mostly cashless economy of subsistence family farms was being supplanted by agricultural capitalism. As the historian Percy W. Bidwell noted, "The changes in agricultural technic and in the social life of the rural folk" during that era were "so great and far-reaching that they may well be called an agricultural revolution."[19]

AMERICA'S FIRST AGRICULTURAL REVOLUTION

MANY MILLENNIAL UTOPIANS BECAME small-scale farmers because they wanted to create an alternative to the industrial food system; they rejected the idea that we must have a system that contributes to environmental degradation and social inequality in order to eat. Instead, they wanted to forge a healthier alternative and show others this new system could transform not just our diets, but also our communities and environment. Similarly, when Thoreau came of age, agricultural capitalism was replacing subsistence agriculture in Concord and much of New England. The young freedom seeker recognized this change made farming and food production more centralized, more monopolistic, less healthful, and less conducive to freedom. He set out to prove that an alternative conducive to freedom was still possible.

Within shouting distance of the North Bridge in Concord, where the American Revolution began when, as Emerson famously wrote in "Concord Hymn," a makeshift army of "embattled farmers stood and fired the shot heard round the world," their progeny continued working family land well into the 1800s. During those decades, the number and size of Concord's farms remained stable, about two hundred farms averaging sixty acres each. Unlike southern plantations, farms in New England didn't specialize; instead, into the nineteenth century "every farmer distributed his land in about the same proportions into pasturage, woodland, and tillage, and raised about the same crops and kept about the same kind and quantity of stock as every other farmer."[20]

Despite the mythology of New England's self-sufficient farmers, farms were rarely sufficient unto themselves. Instead, they were interdependent: Farmers borrowed and bartered, exchanging work and goods, striving to make sure credits and debts roughly balanced at year's end. Storekeepers also facilitated barter; farmers often labored on a shop owner's land in exchange for store-bought essentials such as salt, molasses, liquor, and textiles. By collaborating in this way, farmers eked out a modest living. Lean though their yields often were, they had little incentive to develop new cultivation methods because opportunities to trade or sell surplus were limited. Farmers with extra could sell it on market days in nearby cities and large towns, generating enough cash to pay their taxes. But roads were so poor and farms so widespread that transporting fragile items was impracticable: Milk, for example, was seldom sold more than fifteen miles from the farm.

Reformers had long pressed New England farmers to adopt the methods that enabled their English counterparts in the eighteenth century to increase their yields. But those British bumper crops were generated to meet a very different situation, as the island had far less land and began industrializing earlier. By 1800, more than 20 percent of Great Britain's population lived in cities and had to rely on food grown by others; just 6 percent of Americans did. Then, in 1813, industrialization came to Concord courtesy of Francis Cabot Lowell and his partners, who built a textile mill less than ten miles away. The mill soon moved north, where it was joined by others. Named after Lowell, the city that grew up around them is still known as the "cradle of the American Industrial Revolution."

The region's surge of jobs in manufacturing, construction, and trade created demand for whatever local farmers could bring to market. Soon, competition drove them to grow more and to begin specializing. Concord's soil was good for growing grasses, so many farmers raised cereal grains and pastured cattle—though few put all their eggs in those two baskets.

Because the roads remained poor, markets for perishables continued to be local. But in 1825, the 363-mile Erie Canal was completed, connecting the eastern seaboard's cities and ports to western New York and to the flat, loamy midwestern plains beyond. In principle, the Erie Canal opened new markets for New England farmers. But it tended to hurt them more than it helped, as they were at a disadvantage relative to their peers in other regions. The soil in much of New England was depleted and required a lot of inputs to be fertile. And the region's uneven topography made it impossible to use new laborsaving devices like mowing machines and horse-drawn reapers.

Not quite two decades later, in June 1844, the railroad reached Concord. The hulking iron horses could chug a mile in just two minutes. More than the Erie Canal had, the railroad increased the town's connectivity with the rest of the country and presented farmers with both more competition and more market access. Some balked at the need to change the way they worked; others embraced the new model. Many added marketable items such as oats, hay, and wood to their production mix, much as farmers today add cutting flowers and herbs.

But here's the rub: Even as the farmers grew more product for market, they couldn't stop growing for home use. Rather, as the historian Robert Gross noted, "Farmers were adding greatly to the burdens of their work. One crop was not substituted for another. Farmers simply exploited themselves more intensively than ever." *Exploited* is precisely the right word; the new tasks were more exhausting than the old ones. The farmers used to "spread their labor over the land, plowing shallowly, manuring thinly, and cultivating infrequently, with the result that yields were low," but the changes meant they "depended for a living on far more intensive work: chopping wood, reclaiming land for English hay, digging potatoes, making butter, and occasionally even nursing mulberry bushes." Even excellent farmers strained under the increased workload.[21]

The move from a subsistence-centered, somewhat communal farming ethos to a commercial one didn't happen overnight, but by the end of the 1840s, the new techne prevailed. The increased access to markets prompted farmers to work harder to maximize profits. Collective solutions gave way to individual enterprise; husking bees and apple bees, once seen as congenial ways to complete big tasks, were "condemned as uneconomical and wasteful 'frolics.'" Gross summed up the situation: "An era had come to an end; farmers now relied on the claims of cash rather than the chains of community to do their work." Even the US government sensed that farming was changing. In 1839,

Congress appropriated funds to begin collecting statistical data about American farms and conducted the first national agricultural census the following year.[22]

The new circumstances continued to demand more from farmers. Those who'd expanded their cattle ventures found themselves competing with farmers far afield—and not doing well. Between 1840 and 1854, the proportion of locally raised beef cattle sold in Massachusetts dropped significantly: In 1840, two-thirds of the cattle sold there had also been raised there; in 1854, less than half had been. Little surprise that many Concord farmers turned to dairying. Already well versed in managing cattle, and with urban markets for milk and butter now easier to reach thanks to the train, farmers began to increase their herd sizes. Between 1840 and 1850, the average number in a herd rose from five to six and the percentage of farmers who owned at least ten cows doubled, from 11 percent to 22 percent.[23]

That statistic suggests folks were thriving—and some were. But the inverse was as often the case. In the 1840s, widespread changes in the economy and in agricultural technology doomed many family farms. Secretary of Agriculture Earl Butz's infamous urging, in the 1970s, that farmers "get bigger, get better, or get out" didn't have a literal counterpart in the 1840s, but the market itself made the same demand. And in its wake, many family farms failed.

Some were lost to the bank, as farmers mortgaged properties to expand and then fell behind financially. As Thoreau wrote, many families in Concord had been "toiling twenty, thirty, or forty years" to pay off their mortgages, but few achieved that goal: "The man who has actually paid for his farm with labor on it is so rare that every neighbor can point to him. I doubt if there are three such men in Concord." Some farms were lost to operator failure, for farmers needed an array of new skills. Those who had once needed to worry only about the weather began needing to pay attention to "the condition of the market, transportation rates, the tariff, taxation, labor costs, machinery and fertilizer prices, credit, indebtedness, land values, foreign competition, monopolistic practices of industry" as well as the weather.[24]

Still other farms were lost because the next generation didn't want to stay on the land. Watching their parents struggle through this transitional period, many children rejected the exhaustion-inducing enterprise. In the *New England Farmer*, a writer bemoaned the exodus: "Every farmer's son and daughter are in pursuit of some genteel mode of living. After consuming the farm in the expenses of a fashionable, flashy, fanciful education, they leave the honorable profession of their fathers to become doctors, lawyers, merchants, or ministers or something of the kind."[25]

These agricultural changes didn't only affect the farmers. They constituted a major departure in how people in the United States met a basic need, and both issues mattered to Thoreau. During the nation's early years, folks in Concord—as in other New England communities—were, by and large, directly responsible for feeding themselves and their families, often living without much cash. By the end of the 1840s, almost all the region's farmers exchanged goods for money. This agricultural revolution set Americans on a path of near total dependence on others to eat, as we outsourced the capacity to feed ourselves. The most essential thing we must do—the one task even more basic than finding shelter and clothing—we have pretty much ceded to strangers.

Thoreau balked at this loss of self-sufficiency and pointed out how foolish it was for farmers who grew grain to feed their livestock to buy flour for their families. If only they ate what they grew, he wrote, "every New Englander might easily raise all his own breadstuffs." But instead, "fresh and sweet meal is rarely sold in the shops, and hominy and corn in a still coarser form are hardly used by any. For the most part the farmer gives to his cattle and hogs the grain of his own producing, and buys flour, which is at least no more wholesome, at a greater cost, at the

store." As we'll see, concerns about the quality and wholesomeness of store-bought flour recur in every back-to-the-land era.[26]

I'll be honest: I'm glad I don't have to raise all my food. By
February, I'd be desperate for a meal that didn't contain potatoes,
dried beans, beets, or stewed tomatoes. But like Thoreau, and like
every generation of American back-to-the-landers, I know leaving food production to strangers can have myriad negative consequences, some merely distasteful, others downright terrifying.
We'll look at a few of those later.

THOREAU'S CAPITALIST WOES

FROM HIS SHANTY ON the edge of Walden Pond, Thoreau was
developing a critique of capitalism resonant with one being articulated by Karl Marx and Friedrich Engels, who published their
first coauthored work in 1844. Like Thoreau, Marx believed agricultural capitalism was bad for farmers and for the land: "All
progress in capitalist agriculture is a progress in the art, not only
of robbing the worker, but of robbing the soil; all progress in increasing fertility of the soil for a given time is a progress toward
ruining the more long-lasting sources of fertility." Marx goes on
to specifically decry US practices, writing, "The more a country,
the United States of North America, for instance, develops itself
on the basis of great industry, the more this process of destruction takes place quickly."[27]

Neither Marx's nor Thoreau's critiques of farmers were harangues about the ordinary toil of farm life; they were explicitly
about the negative effects of agricultural capitalism. In fact, Thoreau quite liked subsistence farming, and more than once considered it as an occupation. In his journal, he observed that "the
farmer increases the extent of habitable earth. He makes soil.
That is an honorable occupation." His disdain for agricultural
capitalism, on the other hand, was due partly to the ways it sep-

arates eaters from their food, but mostly to the ways it separates farmers from their own freedom.[28]

Agricultural capitalism demanded so much work that farmers lacked the time and energy to cultivate personal freedom; they became, in Thoreau's opinion, "the tools of their tools." He believed farmers had to focus so much on the "superfluously coarse labors of life that its finer fruits cannot be plucked by them. Their fingers, from excessive toil, are too clumsy and tremble too much for that . . . The laboring man has not leisure for a true integrity day by day; he cannot afford to sustain the manliest relations to men; his labor would be depreciated in the market. He has no time to be anything but a machine." Thoreau's laboring man is the American first cousin to Marx's alienated worker, both reduced to automaton-like activity in the service of capital, both diminished in their sense of their own humanity.[29]

Little wonder that Thoreau saw the situation of textile mill-workers as much the same. He thought the young women choosing the factory over the farm were replacing one form of endless labor with another: "The condition of the operatives is becoming every day more like that of the English, and it cannot be wondered at, since, as far as I have heard or observed, the principal object is, not that mankind may be well and honestly clad, but, unquestionably, that corporations may be enriched."[30]

Because Thoreau understood that capitalism depends on consumerism, he also regarded owning lots of material objects an obstacle to living fully. Grasping the premise of exchange-value, he wrote: "The cost of a thing is the amount of what I will call life which is required to be exchanged for it, immediately or in the long run." He scathingly feigned to wonder whether "the respectable citizen" really needed to teach the next generation to have "a certain number of superfluous glow-shoes and umbrellas, and empty guest chambers for empty guests." Thoreau, no doubt, would have been happy with one pair of glow-shoes.[31]

Not that Thoreau exalted poverty. He regarded it, too, as soul-killing. For much of Thoreau's life, his family scraped by financially. What's more, he graduated from college right as the Panic of 1837 began and many once economically secure families floundered. As bank after bank began to falter, then fail, so too did the economy. Before the recession ended, 353 of the nation's 850 banks—40 percent—had closed. The dollar deflated and unemployment reached 25 percent. During that period, Thoreau held several unsatisfying jobs, and eventually became a handyman for Emerson.

While living at Walden Pond, Thoreau witnessed acute involuntary poverty: "I see in my daily walks human beings living in sties, and all winter with an open door, for the sake of light, without any visible . . . wood pile, and the forms of both old and young are permanently contracted by the long habit of shrinking from cold and misery, and the development of all their limbs and faculties is checked." He began to come to an understanding of poverty that resonates with that of subsequent generations of back-to-the-landers, as he realized "the luxury of one class is counterbalanced by the indigence of another."[32]

POND LIFE

EVEN AS HE PARED down his belongings, Thoreau didn't mirror the conditions of involuntary poverty others living near the pond endured. Simplifying wasn't about getting rid of all material goods; it was about clearing away the inessentials so freedom could flourish. Thoreau's possessions were almost all practical. A replica of his cabin now sits on the site where Thoreau built it. Made with hand-hewn logs, with the bark still evident on the crossbeams' half logs, the small building is solid and beautifully proportioned. The door and fireplace are across from one another on the short sides and a single, large twelve-over-twelve window with rippled panes lets light in on each of the long sides. A narrow bed defines one side as his bedroom; his desk and a table mark the other as his study and kitchen. Other items include three chairs, cooking supplies and eating utensils, a mirror and washbasin, a lamp, some oil, even a cookstove. And, of course, he had notebooks and books, among them not one but two copies of Homer's *Iliad* (one in English, the other in Greek).

Although considerably more expensive than the hovels by the railyard, his cabin was still inexpensive. Not including the lend of the land and ax, the materials he found on site, and his own labor, Thoreau tallied the cost of his cabin as $28.12½, about what a single renter would pay annually to a landlord. Moreover, he built it with little help, tacitly thumbing his nose at capitalism's insistence on labor specialization and proving that a decent shelter need not subjugate its owner.

His foraging for wild foods similarly flouted the emerging capitalist food paradigm, as it enabled him to enjoy free food. Moreover, like all foragers, Thoreau needed to be able to distinguish among edible and toxic plants and mushrooms; he had to be in an intimate rather than a transactional relationship with the wider world in order to safely eat his bounty. He especially loved to gather berries: When he got out of jail after refusing to pay his poll tax, he celebrated his freedom by going huckleberrying. He also gathered cranberries, strawberries, blueberries, and blackberries, and was adamant that the only berries worth eating were those he'd just picked: "Fruits do not yield their true flavor to the purchaser of them, nor to him who raises them for the market. There is but one way to obtain it . . . It is a vulgar error to suppose that you have tasted huckleberries who never plucked them" because "the ambrosial and essential part of the fruit is lost with the bloom which is rubbed off in the market cart, and they become mere provender." Even more than his garden, foraging allowed Thoreau to "live in each season as it passes—breathe the air, drink the drink, taste the fruit."[33]

The large garden he cultivated at Walden Pond was a source of food and pleasure—and a reproof to those who believed agricultural capitalism to be the only way to farm successfully. In "The Bean Field," seemingly the most DIY-ish section of *Walden*, he wrote at length about tending his seven miles of bean: "I came to love my rows, my beans, though so many more than I wanted." He "cherished" them and spent "from five o'clock in the morning till noon" tending to them, coming "to their rescue armed with a hoe [to] thin the ranks of their enemies, filling up the trenches with weedy dead." Despite these efforts, his bean yield was abysmal, half as much as he might reasonably have expected.[34]

More than simply being his market crop, the beans were a set of opportunities. Growing them, Thoreau wrote, "attached me to the earth, and so I got strength." They also gave Thoreau agricultural acumen he could deploy in his writing metaphorically: "Some," he explained, "must work in fields if only for the sake of tropes and expression, to serve a parable-maker one day." As he learned about beans, he also learned about the weeds that accompany them and grew skilled at recognizing and removing cinquefoil, Roman wormwood, pigweed, sorrel, millet grass, and piper grass (I can't help wondering if he ate the sorrel and pigweed with a bit of cinquefoil as a foraged salad). Thoreau also had the sublime experience of losing himself in his work: "It was no longer beans that I hoed, nor I that hoed beans," he wrote. And he sensed, briefly, that this mode of farming connected him both to the Native people whose tools he unearthed while tilling his fields and to the ancients for whom husbandry was "a sacred art."[35]

He contrasted those fine feelings with the worldly efforts of farmers who concentrated on "large farms and large crops merely." Ultimately, he concluded that a farm-centric focus missed the point: "It matters little comparatively whether the fields fill the farmer's barns. The true husbandman will cease from anxiety, as the squirrels manifest no concern whether the woods will bear chestnuts this year or not, and finish his labor with every day, relinquishing all claim to the produce of his fields."[36]

Although that's poor counsel to offer an aspiring homesteader, Thoreau is stressing that soul cultivation is his true priority. He even wrote that he wouldn't "plant beans and corn with so much industry another summer," preferring to sow the seeds of "sincerity, truth, simplicity, faith, innocence, and the like." Those seeds, Thoreau lamented ruefully, must have been "wormeaten or had lost their vitality," as they didn't grow. Nevertheless, he was adamant people would be better off if they concerned themselves less "about our beans for seed" and more about cultivating "a new generation of men" imbued with "truth or justice" or "the kernel of worth and friendliness."[37]

In his reflections on eating fish, Thoreau was similarly both earnest and arch. In the middle section of *Walden*, "Higher Laws," he wrote, "When at the pond, I wished sometimes to add fish to my fare for variety. I have actually fished from the same kind of necessity that the first fishers did. Whatever humanity I might conjure up against it was all factitious, and concerned my philosophy more than my feelings," though he was quick to add, "Not that I am less humane than others, but I did not perceive that my feelings were much affected. I did not pity the fishes nor the worms." Just two pages later, though, he reversed himself: "I have found repeatedly, of late years, that I cannot fish without falling a little in self-respect." By the time he wrote the final draft of *Walden*, he no longer fished at all. But, he wrote, "If I were to live in a wilderness I should be tempted again to become a fisher and hunter in earnest."[38]

This to-ing and fro-ing may sound confusing and contradictory, but it helped him make a larger point. Although often judgmental about individuals, Thoreau believed in the perfectibility of human beings and of human civilization—if we cultivate our finest nature. And not eating other animals was, by his lights, a feature of a finer nature: "Whatever my own practice may be, I have no doubt that it is a part of the destiny of the human race, in its gradual improvement, to leave off eating animals, as surely as the savage tribes have left off eating each other when they came in contact with the more civilized." We'll see a similar emphasis on vegetarianism or vegetarian-adjacency in many of the back-to-the-landers' diets, as most believe we're in relationship not only with other people, but with the rest of the animate world as well.[39]

As would his transcendental peers and successive generations of back-to-the-landers, Thoreau embraced a broad sense of reciprocity with the rest of the living world and saw himself as embedded in an extravagant universe. In the opening of "Solitude," for example, he wrote that his "whole body is one sense, and imbibes delight through every pore. I go and come with a strange liberty in Nature, as part of herself." Rather than being an inchoate oneness, his sensation is grounded in specifics, for he continues, "I walk along the stony shore of the pond in my shirt-sleeves, though it is cool as well as cloudy and windy, and I see nothing special to attract me, all the elements are unusually congenial to me," adding that "the bullfrogs trump to usher in the night, and the note of the whip-poor-will is borne on the rippling wind from over the water. Sympathy with the fluttering alder and poplar leaves almost takes away my breath; yet, like the lake, my serenity is rippled but not ruffled."[40]

Later, Thoreau recalled worrying he might need human neighbors to be happy and healthy. But he was soon reassured, for "while these thoughts prevailed, I was suddenly sensible of such sweet and beneficent society in Nature." He grew conscious of "an infinite and unaccountable friendliness all at once like an atmosphere sustaining me, as made the fancied advantages of human neighborhood insignificant, and I have never thought of them since. Every little pine needle expanded and swelled with sympathy and befriended me." His attunement to the natural world ensured his never being alone: He was always in relationship and in community with his surroundings.[41]

PRAGMATIST IN PARADISE

Thoreau moved to the edge of Walden Pond to live outside an economic system he found pernicious. He hoped this would help him live truly, wisely, freely—that he could live what Scott and Helen Nearing would later call "the good life." One element of that good life, for all the generations of back-to-the-landers, is a welcoming spirit. And although Thoreau famously kept just three chairs—"one for solitude, two for friendship, three for society"—his little cabin became not simply the occasional salon, but also a destination for picnickers and others. People were curious about his life there and stopped by to see him. If he wasn't home, visitors often "left their cards, either a bunch of flowers, or a wreath of evergreen, or a name in pencil on a yellow walnut leaf or a chip." In *Walden Pond: A History*, W. Barksdale Maynard wrote pointedly that Thoreau was obliged to welcome visitors because utopia requires witnesses. Similarly, the Thoreau scholar Laura Walls described his time living at Walden Pond as "an iconic work of performance art."[42]

Thoreau's sociality, though, is one of the things often held against him by critics. Castigated on many fronts, he is often accused of hypocrisy, usually because he went to Concord several times a week and ate dinner with friends and family—and because his mother did his laundry. Such fault-finding began during Thoreau's life, reached what should have been its apotheosis in James Russell Lowell's exquisitely unkind assessment in 1865, and continues today. In the last decade or so, when not *ad hominem*, these critiques have become embedded in a broader appraisal of his social privilege.

Examining the role of privilege in American culture and literature is important; however, framing Thoreau's in terms of dinners out and dirty laundry seems misguided to me. Thoreau could dwell safely and comfortably on the outskirts of town not because his mom fed him or washed his clothes but because he was white, male, single, able-bodied, a local son, a keen observer, a practiced outdoorsman, a skilled forager and gardener, a capable craftsman, and a friend of the landowner. Without any one of those attributes, he'd have had a tougher go of it. But as much as those attributes benefited him, his privilege is most apparent not in his ability to spend time at Walden, but in his ability *to leave.* He could move back into town whenever he wanted.

Not so for the less-privileged others who lived near Walden Pond. The area was then occupied by Irish outcasts and by the ghosts of other, earlier outcasts; both the present and past squatters resided there because it was one of the few areas where Concord's other residents permitted them to live. For forty years, beginning shortly after the American Revolution, a village of fifteen free Blacks lived on the hillside above Walden Pond. Among those whom Thoreau mentions in *Walden* is Brister Freeman, who gained his liberty during the war. Freeman, his wife, Fenda, their three children, and Zilpha White, likely Freeman's sister, squatted near the spot Thoreau turned into a bean field.

Thoreau squatted on land Emerson lent him; even so, I think Elise Lemire is correct in arguing that the reverberations between his efforts to live freely and those of Walden Pond's past Black residents mattered to Thoreau. Lemire proposes that he went to Walden specifically in recognition of, and a kind of solidarity with, the free Blacks who'd lived there, that "it was the marks left on the landscape by the former slaves and other outcasts as much as the plants and animals that made these areas so interesting and thus he devotes a section of *Walden* to their memory." While he lived at Walden, Thoreau certainly became more thoughtful about how his own freedom was intertwined with that of others.

And in the wake of his time there, he became willing to publicly express his abolitionist views.[43]

THOREAU'S ABOLITIONISM

THOREAU CONCENTRATED ON CULTIVATING his own freedom while he lived at Walden Pond, but he gradually expanded his understanding of freedom, grasping its essentially collective nature, after returning to Concord proper and becoming more involved with the abolitionist activities his relatives pursued. The family home was a station on the Underground Railroad, and six of the eight women who founded the Concord Women's Antislavery Society—Thoreau's mother, his sisters Helen and Sophia, and three aunts (Maria Thoreau, Jane Thoreau, and Louisa Dunbar)—lived there. The Concord chapter was well organized, provided money to many other antislavery groups, and operated as a haven for runaway slaves heading north. Helen Thoreau was friends with Frederick Douglass and William Lloyd Garrison.

Henry David Thoreau had for a while played a less prominent role in the abolitionist movement. He did help runaways on their northward journeys by guiding them to the next safe house or buying them train tickets and boarding with them to make sure they were on their way. Several spent time with him at Walden Pond, waiting for darkness, before continuing north. And at his tiny cabin, Thoreau hosted a First of August celebration, commemorating the emancipation of slaves in the West Indies. Although he was the host, Thoreau played a behind-the-scenes role. Others gave the speeches; he rang the meetinghouse bell to announce the start of the festivities.[44]

Gradually, Thoreau moved toward publicly condemning slavery. In April 1851, Thomas Sims was arrested in Boston as a fugitive slave. Thoreau railed against the arrest in his journal, but not in public. He wrote that he couldn't trust the government to do the right thing, that he "would not trust the life of my friends to the judges of all the Supreme Courts in the world put together." Finally, in 1854, when the state of Massachusetts returned the fugitive slave Anthony Burns to Virginia, Thoreau turned the corner. Again convinced of the government's perfidy in ignoring its citizens' opposition to returning Burns, he spoke out.[45]

Thoreau's public support of abolition helped extend the movement's reach, as he was by then an established public intellectual; he used that stature—his platform, as it were—to condemn slavery. At a rally organized by the Massachusetts Anti-Slavery Society on the Fourth of July, standing beneath an American flag hung upside down, Thoreau shared a stage with, among others, Sojourner Truth. He denounced the Massachusetts judiciary for its support of the Fugitive Slave Act: "I have lived for the last month—and I think that every man in Massachusetts capable of the sentiment of patriotism must have had a similar experience—with the sense of having suffered a vast and indefinite loss. I did not know at first what ailed me. At last it occurred to me that what I had lost was a country." As he had years earlier in "Civil Disobedience," he rejected the authority of government when it behaved unjustly: "Show me a free state, and a court truly of justice, and I will fight for them, if need be; but show me Massachusetts, and I refuse her my allegiance, and express contempt for her courts." He reiterated what Orestes Brownson had laid out almost twenty years earlier—that freedom for one is possible only if there's freedom for all.[46]

After John Brown was captured by the Marines following the Harpers Ferry arsenal raid, Thoreau raged in his journal, calling the government "a demoniacal force!" and one "that pretends to be Christian and crucifies a million Christs every day!" Amid the fury-filled invectives about the government, politicians, and toadying news editors, however, he described John Brown as a hero using the language of a plants-man: "In the moral world, when good seed is planted, good fruit is inevitable and does not depend on our watering and cultivating; that when you plant, or bury, a hero in this field, a crop of heroes is sure to spring up. This is a seed of such force and vitality that it does not ask our leave to germinate." John Brown, this rhetoric suggests, was a member of the "new generation of men" he'd written about in *Walden*, the one the world sorely needed. Indeed, during the month-plus between the raid and Brown's execution, Thoreau on several occasions delivered a lecture entitled "A Plea for Captain John Brown," hoping to convince people he wasn't a fool, that, to the contrary, Brown was a patriot "like the best of those who stood at our bridge once, on Lexington Common, and on Bunker Hill."[47]

Such social-justice concerns are part of all five of the utopian back-to-the-land movements discussed in this book. The idealists often engaged the most pressing issues of their day. From slavery to civil rights, from gender equality to economic equality, from food justice to climate justice, they sought a freer, more egalitarian social order not just for themselves, but for all. Like Thoreau, most of the people in these movements were white, able-bodied, and educated. And, again like Thoreau, many were conscious of their privilege and concertedly used it to further social-justice aims. They recognized their freedom as inextricably tied to that of others, just as their life on the land connected them to vast ecosystems of beings with whom they inhabited the place.

BEYOND THE POND

IN THE FIFTEEN YEARS between leaving Walden Pond and his death, at age forty-four, from tuberculosis, Thoreau split his time between pursuing paid labor and his true vocations. He worked primarily as a surveyor, staking out property boundaries. He was careful about how he apportioned his time; when possible, he allocated morning to paid work and the rest of the day to his interests. Likely, being able to work outdoors, by himself, on his own schedule made more tolerable the fact that surveys are used to turn land into a commodity.

By the mid-1850s, Thoreau no longer imagined one could live self-sufficiently. He still sometimes fantasized about going "off to some wilderness where I can have a better opportunity to play life," but forswore such possibilities, asking in his journal, "What is the use of trying to live simply, raising what you eat, making what you wear, building what you inhabit, burning what you cut or dig, when those to whom you are allied insanely want and will have a thousand other things which neither you nor they can raise and nobody else, perchance, will pay for?" After all, he points out, "The fellow-man to whom you are yoked is a steer that is ever bolting right the other way."

Yoked, yes, but not solely to others with a thousand wants and not every minute of every day. By dwelling among family and like-minded friends and by saving afternoons for himself, Thoreau was able to live modestly, albeit not simply, and to maintain an intimate relationship with the natural world. As discouraged as he sounds in that journal entry, he finished the day—and the entry—with detailed descriptions of all he saw during an afternoon walk in Depot Field Brook, Hubbard's Crossing, Potter's Swamp, and Cardinal Shore on his way to Fair Haven Hill.[48]

Despite those expressions of uncertainty and frustration, Henry David Thoreau remains one of America's most famous seekers of freedom, lovers of nature, and advocates of simple living. Many others of his era aspired to live by those values, and also imagined doing so would be easier outside the constraints of culture. They, too, tried to realize nether paradises, heavens on earth. As we'll see, some of their efforts contained inadvertently hilarious elements. But at their core, all dreamed of creating communities where people could be free and live in harmony.

3

SECULAR UTOPIAS OF THE 1840s

Before Thoreau headed to Walden Pond for his experiment in simple living, he'd had opportunities to participate in communal efforts. George Ripley invited him to join Brook Farm, the utopian community he and his wife founded in West Roxbury, Massachusetts, and Bronson Alcott asked him to join his family and the others creating a "new Eden" at Fruitlands, in Harvard, Massachusetts. An emphatic non-joiner, Thoreau said no to both friends, though he visited them often. In addition to Brook Farm and Fruitlands, the Bay State was then home to five Shaker communities, two of them in easy walking distance from Concord (easy for Thoreau, that is), as well as to the Hopedale Community and the Northampton Association for Education and Industry, where Abigail Alcott's brother lived.

Other experimental communities—some eighty during the 1840s—were scattered throughout New England and the Midwest and attracted tens of thousands of participants. Although virtually all of these utopian groups sought a greater sense of freedom, equality, the abolition of slavery, and communal sufficiency, not all were equally committed to living off the land. Thoreau's effort at Walden Pond and Alcott's at Fruitlands most fully embodied the belief that self-sufficiency is necessary to true freedom. Otherwise, people must interact with organizations or systems they find unjust in order to eat, or find shelter, or earn money. Hopedale and Northampton both remained economically connected to the mainstream culture; they ran manufacturing businesses. Likewise, the Shakers sold beautiful handmade goods. Brook Farm ran a school open to children of both members and outsiders.

Isaac Hecker, the friend who'd invited Thoreau on a walking tour of Europe, was a resident first at Brook Farm and then at Fruitlands. Like him, many other seekers moved among the various experimental communities, cross-pollinating them as folks "visited, corresponded, and in other ways influenced each other." These seekers, like utopians more generally, regarded mainstream society as in decline, failing to live up to the promise of freedom and equality that had animated the nation's founding. Participants believed that if only they modeled better alternatives, others would see their value and replicate them.[1]

Despite their shared beliefs and ample cross-pollination, the communities were far from interchangeable. Both the Shakers and the Oneida Community strove for gender equality, for example, with the Shakers requiring celibacy in part to attain that aim. In contrast, the Oneida Community believed "complex marriage" would help bring about equality. Instead of being celibate or sexually exclusive, adult members were encouraged to have sex with any other, and as many other, consenting adults as they wanted to.

Just as the Shakers and Oneidans took opposite tacks to reach the same end, Fruitlands and Brook Farm, the two communities established by transcendentalists, followed quite different paths in their efforts to achieve freedom and equality and regain an awareness of the divinity within. One was ascetic, the other trended toward pleasure-seeking; one was mostly celibate, the other kindled more than a dozen marriages. One centered on individual betterment first, the other on reorganizing social institutions first. The two were also distinct in their food rules, though they both tied food production and consumption to the quest for freedom. Neither Fruitlands nor Brook Farm succeeded for long, as each was hampered by conflict, bad luck, and poor planning. Even so, they're regarded as two of the most important utopian experiments of their era, in no small part because they were among the very few secular utopias.

ALCOTT'S NEW EDEN

THIRTY MILES WEST OF Boston, in Harvard, Massachusetts, Prospect Hill Road follows the contour of the Nashua River Valley, providing beautiful views of fields and woodlands below and Mount Wachusett beyond. From there, you can also see Fruitlands, once Amos Bronson Alcott's farm. The modest clapboard house has been well maintained of late; it's in far better shape now than when the hopeful farmers lived there. When they moved in, the kitchen was in a shambles, the chimney was unusable, and part of the roof had caved in. Now, the interior is clean and well appointed, and the exterior is a bold red that stands out nicely against both summer greens and winter whites.[2]

Alcott had been inspired by Brook Farm, then two years old, though he didn't think the utopians there were quite ascetic enough. He envisioned creating a much leaner, more austere "New Eden" where all beings could live in harmony with one another, fully self-sufficient and separate from the sullying influences of the larger world. Freed from the negative effects of capitalism and the increasingly industrial culture, each resident would be able to access the divine within and thus together they could establish a miniature paradise. Accompanying Bronson Alcott in creating this "Consociate Family" were his wife, Abigail, and their four young daughters, Anna, Louisa May, Elizabeth ("Lizzie"), and Abby May; Charles Lane and his son, William; and a small, changing group of acolytes.

The core group moved to Fruitlands on an unseasonably chilly June 1 in 1843. Given their desire to grow all their own food, they really should have had most of it planted by then. But only a few of them had experience tending anything bigger than a kitchen garden, and just one had significant agricultural abilities. That real farmer—a middle-aged man named Joseph Palmer—was thwarted by the others' idiosyncratic ideas. They refused to use manure to enrich the soil, for example, because they thought it would defile the soil's purity and because they thought it exploited the animals who produced the manure (yeah, I don't know why that's exploitative, either). The destiny of not-particularly-good farmers trying to grow all their food in not-particularly-good soil is not particularly surprising: Underfed, with nothing to put by for winter, the group fell apart in less than a year.

Had this been an ordinary farm, it would almost certainly have suffered the same fate as have most small farms in that area: The ninety acres of prime real estate would now be a suburban housing tract whose promoters tout the advantages of country life combined with easy access to the Mass Pike and Boston. Instead, despite its failure, the farmhouse and surrounding land have been preserved as a National Historic Landmark, in honor of the utopian experiment.

I grew up about ten miles from Fruitlands and have visited the farm-turned-museum many times. One early memory is of a grade school class trip to see the place Louisa May Alcott lived. Years later, when I was an English major in college, friends and I made the requisite Thoreauvian pilgrimage to Walden Pond, detouring to nearby Harvard so I could show them where one of the other transcendentalists had lived. As far as I was concerned back then, Amos Bronson Alcott's claims to fame were that he was Louisa May's dad and a lesser-known transcendentalist. Recently, I returned to Fruitlands to find out more, to learn why he moved there in his quest for utopia, and to understand why it went so wrong so quickly.

That visit took place on a sweltering August day, the heat index well over a hundred degrees. The attic bedroom where the Alcott sisters slept felt so dangerously claustrophobic that I left after a minute. Louisa May's weirdly enthusiastic diary entry from September 1—"I love cold water!"—suddenly made sense. The rooms below were marginally cooler. One is filled with dusty old hardcover books to connote Alcott and Lane's impressive library. A bust of Socrates, a facsimile of the one Alcott brought with him wherever he moved, now sits beside one of Alcott himself, a gently cleft chin and bushy eyebrows dominating his other features. A couch that belonged to Nathaniel Hawthorne and a writing desk used by Alcott's dear friend and acolyte Henry David Thoreau hold pride of place in other rooms.[3]

Bronson Alcott didn't initially aspire to be a self-sufficient farmer living outside the mainstream; he wanted to be a teacher. The erratic course of his teaching career led him to this experiment in simple living—which helps explain why little about the house suggests it had been anything other than the home of an unusually literate family. Alcott's agricultural emendations have not survived: Planted too close to the building, his mulberry trees undermined the foundation and had to be removed. The fields have mostly returned to woods. But the story of Bronson Alcott's time there has been well preserved.

Like the other transcendentalists, Bronson Alcott believed humans were innately divine and could be more fully in tune with the spiritual realm while still alive—if only they could figure out how. Children, he reasoned, retained greater access to their innate divinity than did adults as they'd been separated from it for a shorter time. As he searched for techniques to help them tap into the divine, he cultivated novel teaching methods that earned wide attention. Financial backers who admired his ideas helped Alcott open schools in Connecticut, Pennsylvania, and Massachusetts. But sooner or later, Bronson's eccentricity would shock the parents of his students, and he'd be let go or the school closed.

Ralph Waldo Emerson was also impressed by Alcott's ideas. The two met in October 1835 and hit it off right away. Alcott recorded this in his journal: "I have not found a man in whose whole mind I felt more sympathy than in his." Emerson urged Alcott to move to Concord so he could become more fully part of the intellectual community coalescing there. In 1840, after Alcott's Temple School in Boston closed, the Alcott family moved to Dove Cottage, a short walk from Emerson's home. Disappointment over the school's failure added to Alcott's growing disgust with cities, which, he decided, were "hostile to liberal learning . . . A suburban neighborhood at least, if not a rural, were most friendly and inviting."[4]

An excellent conversationalist and an important contributor to transcendentalism, Alcott was not a very good writer. But Emerson admired Alcott's ideas so much that he volunteered to refine his friend's texts—a kindness undone when Alcott invariably did one more edit. Another of Emerson's generous acts yielded more: He covered Alcott's expenses so he could go to England and meet James Pierrepont Greaves, a reformer and mystic who had found Alcott's approach so compelling that he and several friends had established Alcott House, a school and utopian community. Greaves died before Alcott reached England, but Bronson did visit Alcott House, where he met with Charles Lane, a cofounder, and Henry Gardner Wright, a young teacher. Wright greatly impressed Alcott, and the American begged him to come to the United States so they could work together. On that same trip, Emerson also arranged for Alcott to meet one of his dearest friends, the philosopher and historian Thomas Carlyle. Unfortunately, Carlyle said he and Alcott got along "almost as ill as it was possible for two honest men kindly affected towards one another to do." He was dumbfounded by Alcott's ardent vegetarianism and described him as trying to save the world "by return to acorns and the golden rule."[5]

Alcott persuaded Henry Wright and Charles Lane to come to America to create another Alcott House, in Concord. He believed America better suited than England because, as he wrote, "our freer, but yet far from freed land is the asylum, if asylum there be, for the hope of man: and there, if anywhere, is that second Eden to be planted." As this "divine seed" grew, it would "bruise the head of Evil, and restore Man to his rightful communion with God, in the Paradise of the Good, where-into neither the knowledge of Death nor Sin shall enter; but Life and Immortality shall then come to light, and man pluck wisdom from the tree of life always" (hence, Emerson's efforts to smooth out Alcott's writing). Although his prose was wrought, Alcott's point was widely taken: The young nation was ripe for—and rife with—utopian experiments.[6]

In October, the Lanes and Wright settled into the Concord cottage the Alcotts were renting. By the next spring, when Lane purchased the ninety-acre farm that would become Fruitlands, their plan for a new, improved version of Alcott House had changed dramatically. Mostly due to Lane, the vision was becoming more ascetic and less centered on children. Wright thought the new plan impractical and predicted they'd all freeze or starve to death if they adopted it. Soon, Lane and Wright were squabbling incessantly. Wright was more of an aesthete than Alcott had realized; Lane, more austere and judgmental. Their battle came to a head one morning at breakfast: Wright poured cream on his oatmeal, which enraged Lane, who tried to hit him. Alcott took Lane's side: "For one mortal to threaten another with physical violence was, of course, wrong, but not so wicked as for a grown man to deprive a helpless little calf of the very best of its mother's milk. Mr. Wright forthwith left Mr. Alcott's house."[7]

Though a curmudgeon, Lane agreed with Alcott's vision of Fruitlands as a community guided by love and freedom. As had Thoreau, they thought freedom involved living self-sufficiently and avoiding an economic system that exploited people, animals, and natural resources. To achieve that freedom, the Consociate Family planned to grow most of their food as well as some flax to make linen for clothing. They conceded that they'd buy what they couldn't grow, but the list of allowed items was short. Many of the things other people in New England typically purchased—alcohol, tea, coffee, molasses, rice, dairy products, meats, leather, cotton, and wool, for example—were forbidden because their production exploited people or animals. Molasses, rice, and cotton depended on slave labor; leather and wool exploited animals. In addition, they believed tea and coffee took advantage of those who tended the plants and harmed those who drank the beverages.

Confident about their course, Alcott wrote a letter to a transcendentalist journal, *The Dial*, just ten days after they'd established themselves at Fruitlands. In it, he explained that "ordinary secular farming is not our object"; they were farming to live "in harmony with the primitive instincts in man." They were cultivating land to cultivate spirit: "Fruit, grain, pulse, garden plants and herbs, flax and other vegetable products for food, raiment, and domestic uses, receiving assiduous attention, afford at once ample manual occupation, and chaste supplies for the bodily needs. Consecrated to human freedom, the land awaits the sober culture of devout men." The land did not need to wait long; a few days after arriving, the devout men began planting the eleven acres with corn, barley, rye, beans, potatoes, squashes, melons, and other crops.[8]

"ABSTAIN, ABSTAIN, ABSTAIN!"

AFTER LETTING US ROAM around the farmhouse on our own for a half hour, the docent at the Fruitlands Museum called all the visitors to the kitchen for a question-and-answer session. We were a disparate group—an older couple from Florida, three women from Germany, a little boy (fidgety and clearly bored out of his mind) who must have been the son of one of them, and me. The docent seemed proud of the Alcotts, impressed by their effort, and eager to tell us everything she knew. She grew visibly annoyed when I asked about Alcott's ineffectiveness as a farmer. I didn't want to be *that guy*, the one who tries to show up the tour guide, so I refrained from asking anything more. Happily, she soon moved on to something I'd been wondering about: why Alcott and Lane and so many of that era's other utopians were teetotalers.

Earlier in the nineteenth century, Americans drank a lot of alcohol. And not just hard cider. According to the docent, by 1830 the average American drank more than seven gallons of distilled liquor per year. I did the math when I got home; that's about six hundred shots each. To me, that didn't seem terrible, as most doctors say one drink a day is fine, and six hundred shots a year averages out to 1.6 shots a day. But their distilled liquor wasn't like most of what's available now. Corn based and extremely potent, the only commercial product I know of that's comparable is Everclear, the 190-proof grain alcohol beloved by many a college fraternity and banned in fourteen states.

As it happens, America's drinking problem was quite a bit worse than the docent's description suggested. The seven-gallon average she cited didn't take into account that children drank relatively little alcohol and women imbibed much of theirs in the form of patent medicines—which enabled them to appear more restrained than they actually were. The American Temperance Society estimated that during the 1820s, the three million men in the United States collectively drank sixty million gallons of hard liquor each year, which averages out to seven ounces of corn liquor per man, per day. Every single day. In addition to the hard cider. As the strength and quantity of the alcohol consumed increased, so, too, did "dipsomania." State legislatures began restricting alcohol sales in 1838, making the 1840s a more abstemious period overall.[9]

Like Alcott, most of that era's other utopians were temperance allies who believed drinking alcohol interfered with maintaining physical and spiritual health. Alcott thought alcohol (along with coffee and tea) produced a false sense of mental stimulation that obscured true awareness. Similarly, the Alcotts believed certain foods could enhance or detract from bodily and spiritual purity. After Bronson's cousin William Alcott, a physician, became convinced that vegetarianism was both healthier than a more varied diet and more spiritually pure, Bronson embraced this diet for himself, his wife and daughters, and all the other residents of Fruitlands. At just ten years old, Louisa May saw avoiding meat as a moral issue with far-reaching implications: She wrote in her journal: "without flesh diet / there could be no / blood-shedding war." In 1850, William Alcott would become one of the founders of the American Vegetarian Society, an organization that explicitly connected vegetarianism and purity: "Vegetarianism unfolds the universal law of man's being. Its observance is a stepping-stone to a higher stage of existence, and removes obstruction which hinders the fulfillment of man's highest aspirations, and it is the inlet to a new and holier life."[10]

Another of the society's founders, Sylvester Graham, also significantly influenced Fruitlands' food regimen. Graham's name is now mostly associated with the cracker he introduced, but in his day, he was well known as a crusader who sought to save humankind by promoting temperance. He urged abstaining from not only alcohol, but also meats, spices, fats, and white bread; all, he said, contributed to debauchery. The diet for which he believed "God constructed man" comprised raw fruits, vegetables, and bread made with unbolted flour (flour in which the bran and germ are retained). Moreover, Graham believed that if anyone "disregards these laws, and by artificial means greatly departs from the natural adaption of things, he inevitably brings evil on himself and on his posterity."[11]

Regarding physical and moral well-being as intertwined led Graham to highlight hygiene, too. He advocated bathing more, spending more time outdoors, and exercising. And, anticipating *Dr. Strangelove* by more than a century, he believed people should safeguard their precious bodily fluids; having sex often was dangerous, he warned, as one lost a little bit of "life force" each time. His dietary plan, he assured his followers, would quell carnal passions: Avoiding meat and spices reduced sexual excitability and consuming plenty of whole-wheat bread suppressed the urge to masturbate. I'm skeptical about several of these claims, but Graham's ideas were embraced by so many in his day that his followers became known as "Grahamites." And though I'm especially dubious about his assertion that omnivores bring evil onto their posterity, it's of a piece with claims made by many other utopians I discuss in this book who also believed eating or avoiding certain foods carried karmic consequences.

The central role that whole-wheat bread played in his health regimen prompted Graham to be highly critical of farmers and bakers who didn't strive to make a great product. Like Alcott, Graham believed compost and manure would "debauch" soil, and wanted wheat grown only on "virgin soil." He faulted farmers for forsaking quality for quantity and chided consumers for not demanding better. Similarly, he accused commercial bakers of adulterating flour with "alum, sulphate of zinc, sub-carbonate of magnesium, sub-carbonate of ammonia, [and] sulphate of copper" to hide defects or rancidness, and of adding "chalk, pipe clay and plaster of Paris" to make the flour look whiter—as some did. And, he claimed, even if "public bakers always use the best of flour, their bread, as a general statement, would still be very inferior to well-made domestic bread, in point of sweetness and wholesomeness," so people should make their own.[12]

That last may sound nostalgic, but Graham was right. Then, as now, conventionally milled flours are made with only one part of the wheat kernel, the endosperm. The rest, the high-fiber bran and the nutrient-rich germ, is removed. Thus, even good wheat milled that way generates a less nutritious flour than does the unbolted version. But unbolted flour is so coarse the gluten strands that make bread dough rise are cut by the flour itself during kneading—leading to very dense bread. To make his loaves palatable to his daughters, most of whom still had baby teeth, Bronson Alcott formed them "into the shapes of animals and other pleasant images."[13]

FRUITLANDS' EIGHTFOLD PATH

THE PURE DIET THE Consociate Family followed was to be complemented by "plain garments, pure bathing, unsullied dwellings, open conduct, gentle behavior, kindly sympathies, [and] serene minds." Charles Lane designed the plain garments—loose linen trousers and tunics for the males, bloomers and tunics for the females. The emphasis on pure bathing and unsullied dwellings came from Graham's regimen. But adhering to the other tenets proved difficult. "Serene minds" were hard to reconcile with "open conduct," for Mr. Lane and Mrs. Alcott were often at odds. Tensions began even before they moved to Fruitlands: Mr. Lane imposed his preferences on his host family, didn't help with the daily work, and was, frankly, kind of a jerk. Lane tried to tamp down the joviality of Mrs. Alcott and her daughters, for, as the literary critic John Matteson wrote, he was "utterly without humor and wholly intolerant of frolicsome natures." Tensions worsened when Lane persisted in advocating celibacy as necessary for universal love to bloom—a stricture directed at Bronson and Abigail, Fruitland's sole couple.[14]

Louisa May Alcott seems to have disliked Lane almost as much as her mother did. She recollected her time at Fruitlands vividly, depicting it with satiric insight in a very thinly veiled roman à clef, "Transcendental Wild Oats." In the short story, Charles Lane is "Timon Lion" (sometimes "Dictator Lion"), an egoist who "intended to found a colony of Latter Day Saints, who, under his patriarchal sway, should regenerate the world and glorify his name forever."[15]

On the first evening at the fictional Fruitlands, Timon Lion describes their daily schedule to the group: "We shall rise at dawn, begin the day by bathing, followed by music, and then a chaste repast of fruit and bread. Each one finds congenial occupation till the meridian meal; when some deep-searching conversation gives rest to the body and development to the mind." Likewise, after lunch "healthful labor again engages us till the last meal, when we assemble in social communion, prolonged till sunset, when we retire to sweet repose." Lion refrains from proposing precisely what those congenial occupations will be, as he thinks it essential to avoid "willful activity, which is a check to all divine growth." Perhaps. But "willful activity" is important in agriculture, even when ordinary secular farming has been replaced with soul-cultivation farming. Without it, as the Consociates soon learned, farmers starve.[16]

As Louisa May remembered it, other un-farmer-like ideas and actions also hampered the harvest. The three men sowing grain discovered that "when about half through the job that each had been sowing a different sort of grain in the same field; a mistake which caused much perplexity, as it could not be remedied." Later, when the barley was ready, no men were there to bring it in; Alcott and Lane were out scouting for new members, and the other men had all moved out for good. With a dark storm visible on the horizon, "Mrs. Alcott made a quick decision. Gathering all the baskets she could find, she carried them to the barley-field with the help of the children, and in hot haste they gathered the barley into the baskets and dragged them to the granary, and then ran back as fast as they could for more." They were able to save only enough to last a few weeks.[17]

Still more ill-fated were their orcharding efforts: In midsummer, when the group had settled in, "new trees and vines [were] set, regardless of the unfit season and entire ignorance of the husbandmen, who honestly believed that in the autumn they would reap a bounteous harvest." Whether or not they "honestly believed" they'd have fruit that year, they wouldn't. Mulberry trees don't bear fruit for two or three years. Apple trees take longer, perhaps five years, before they bear good fruit; grapevines take longer still.[18]

Though Bronson Alcott had expected a better harvest than they reaped, he also intended for the Consociates to follow an austere diet. Like William Alcott and Sylvester Graham, he saw physical and spiritual purification as tightly intertwined. The body was both a temple and a tool, an outward expression of spirit and the means for gaining access to the Unity. In "Transcendental Wild Oats," Abel Lamb (Bronson Alcott) enumerates the many, many things to be avoided lest they sully body or spirit: "Shall I stimulate with tea, coffee, or wine! No. Shall I consume flesh! Not if I value health. Shall I subjugate cattle! Shall I claim property in any created thing! Shall I trade! Shall I adopt a form of religion! Shall I interest myself in politics! To how many of these questions . . . would the response be 'Abstain?'"[19]

Louisa May wasn't exaggerating with this litany. She drew it from the longer list Alcott and Lane included in letters to the *Herald of Freedom* and *New York Tribune* touting their plan for crossing "the ocean of life from the narrow island of selfishness to the broad continent of universal love." Voyagers to the broad continent also had to give up milk, warm bathwater, lamps, candles, and layered clothing.[20]

Given the spartan conditions, it's little surprise that few others wanted to live at Fruitlands, though most of the transcendentalists were curious about how the Consociates were faring. When Emerson visited, in July, he was stunned. Despite his frequent financial generosity to Alcott, he'd refused to help fund this venture; he'd written in his journal that he "would as soon exert myself to collect money for a madman." But after visiting, he wrote in his journal, "The sun and the evening sky do not look calmer than Alcott and his family at Fruitlands. They seemed to have arrived at the fact—to have got rid of the show, and so to be serene. Their manners and behavior in the house and the field were those of superior men—of men at rest." He added that "young men and young maidens, old men and women, should visit them and be inspired." Many young men and maidens did drop by. The Fruitlanders welcomed not only their friends, but also "well-wishers, scoffers, and the just curious" from as far away as Ohio and North Carolina. The abolitionist Sojourner Truth heard such good things about Fruitlands that she considered spending the winter there.[21]

Though impressed, Emerson couldn't shake his skepticism. His enthusiastic journal entry ended this way: "They look well in July; we will see them in December." He wouldn't see many. The acolytes, who never numbered more than twelve at a time, were all departed by late November; Charles and William Lane left soon thereafter to join a nearby Shaker community. Lane asked Alcott—just Bronson Alcott, not his family—to go with them. Abigail threatened to move out and take the children if he did; her brother, who was part of the Northampton Association for Education and Industry, got permission for his sister and nieces to join it.

Alcott seems to have struggled with the slowly growing awareness that his ideas about universal love were at odds with his specific love for his wife and daughters. Over Christmas week he went—alone—to Boston; there, he talked with George Ripley about moving to Brook Farm. Ultimately, Alcott decided against it, and also opted not to join the Shakers with Lane. He and his family moved in with a neighbor until they could afford to make a fresh start. After just seven months, the experiment at Fruitlands was over.

BROOK FARM

TWO YEARS BEFORE ALCOTT began Fruitlands, fellow transcendentalists George and Sophia Ripley established Brook Farm on a former dairy farm in West Roxbury, a Boston neighborhood on the city's southwest side. Troubled by the growing social and economic inequality in Boston and the complacency of his congregation, George Ripley left his ministerial post in search of a fresh way to address those problems. In 1840, he attended the Christian Union Convention, in Groton, Massachusetts, where he became inspired to create a utopian community that would be "nothing less than Heaven on Earth." His aim, as he explained in a letter to Emerson a few months later, was "to insure a more natural union between intellectual and manual labor than now exists; to combine the thinker and the worker, as far as possible, in the same person; to guarantee the highest mental freedom, by providing all with labor, adapted to their tastes and talents, and securing to them the fruits of their industry." Agriculture would be their focus, though he also hoped to have a "school or college" associated with it.[22]

Ripley didn't exactly beg Emerson, who was his cousin and close friend, to join Brook Farm, but he made clear how much he wanted him to support it—with both his imprimatur and his wallet. Emerson said no. As did Bronson Alcott, Henry David Thoreau, and Margaret Fuller, despite their fondness for the Ripleys. Alcott wanted to create his own Eden, and Thoreau and Fuller each believed the individual's pursuit of a purer life must precede any communal efforts to create one. Nonetheless, the Ripleys forged ahead.

During the winter, Ripley pored over agricultural periodicals and tried to teach himself about farming, and in late March 1841, he and Sophia and a few others moved to Ellis Farm, a beautiful, 170-acre property abutting the Charles River on one side, with a brook, large open fields, wooded areas, and a low, classic New England stone wall on the side of the property fronting Baker Street. A big barn and two-and-a-half-story colonial-era farmhouse stood mostly ready when they arrived. The group also bought a small adjacent property. Soon, the whole was renamed Brook Farm, and the farmhouse, their center of activity, became known as "the Hive." An ell attached to the Hive contained the laundry room, kitchen, nursery, and a dining room big enough to accommodate as many as fifty people at a time.

Most of the early members espoused the transcendental idea of the divinity of mankind, and from that unfolded a suite of socially minded precepts: a commitment to dissolving class distinctions, a belief in the universal equality of men and women, the conviction that slavery should be abolished. Unlike the Fruitlanders, the Brook Farmers didn't aspire to live without money. Rather, they wanted it distributed more evenly so income inequality wouldn't lead to other inequalities. Brook Farm began as a joint-stock company; each member paid to join, did an equal share of the work, and (it was hoped) reaped a share of the profits. Except for the most pleasant and most unpleasant tasks, all the jobs at Brook Farm paid the same wage. Those who didn't work paid room and board, as did visitors. Even with the initial money paid in and the income from visitors, Brook Farm began—and remained—on shaky financial terrain.

Their economic troubles began with the soil, which was rocky and ill-suited to vegetable cultivation—even after the farmworkers added wagons full of manure from the "gold mine," as Ripley called it. They were exacerbated by general inexperience; as at Fruitlands, only one of the early residents at Brook Farm had much agricultural knowledge. Although eager to learn, the others—many of them members of Boston's intellectual and literary elite—were unfamiliar not just with farm labor, but with manual labor in general. They learned to tend vegetables, milk cows, and care for chickens, but couldn't bear to kill and butcher large livestock. This reluctance led them to question the ethics of eating meat; some did abstain, others continued to eat meat while fretting about the ethics of that choice. Their woes worsened due to the community's apparent success: By the summer of 1842, Brook Farm had seventy residents. Of those, however, fifty-five were boarders, not members, so they didn't have to work. The fifteen who were members scrambled to provide enough food and construct enough living space for all.

That year, George Ripley began learning more about a new philosophy, Fourierism. Developed by François-Marie-Charles Fourier, it was one of the many utopian socialisms to emerge in France in the wake of that country's revolution. In 1799, after witnessing warehouse owners allowing food to rot rather than letting the famine-starved have access to it, Fourier "crystalized the conceptions that had long been floating in his mind," and articulated a social model he believed could ensure everyone's happiness.[23]

Fourier believed that capitalism should be replaced with something more cooperative, that commerce diminished producers and consumers, that society needed to undo class and race and gender inequalities, and that access to education, food, housing, and health should be rights. Though relatively obscure today, Fourierism was highly influential in the 1840s; it even inspired the far-better-known socialists Karl Marx and Friedrich Engels. Marx found Fourier's thinking "profound" and "used it to some extent as a model for the Manifesto of the Communist Party." Engels dismissed Fourierism, but not Fourier himself. Rather, he credited him with identifying "the great axiom of social philosophy"—the notion that most people want to work; they just want jobs they like. If they could have that, people would work "without the intimidation and bribery used by the capitalist system."[24]

Thanks to Albert Brisbane, an ardent proselytizer, Fourierism found a receptive audience in the United States. Brisbane framed Fourier's ideas in terms like those in Orestes Brownson's *New Views*, explaining they would enable this generation to "consummate the great work of Reform commenced by their noble ancestors." Those ancestors had emphasized the centrality of liberty and equality to America's political life, but the time had come to extend them "to its social systems . . . we must effect a Social reform and bequeath to our posterity the far more precious boon of *social* liberty and equality!" Knowing his audience, Brisbane omitted from his account Fourier's most salacious ideas, making the American version far less shocking and radical than the original (though it did retain some fantastical bits). Brisbane's sanitized version resonated so widely that John Humphrey Noyes, who founded the Oneida Community, griped that Fourierism was "drawing into its vortex all the best and leading minds of this country." In less than a decade, forty-five Fourierist communities, or Phalanxes, were established in the United States and the social philosophy had more than one hundred thousand supporters.[25]

George Ripley was initially attracted to Fourierism because of the commonalities between transcendental ideals and the Frenchman's notions. Both held as central the importance of doing manual and mental work; both touted cooperative labor; both wanted to get rid of social and economic inequality; both regarded people as fundamentally interconnected to one another; and both saw people as part of an indivisible "Spirit." Moreover, Fourier's model proposed communal agricultural units organized as joint-stock ventures—precisely what Brook Farm was. Even so, Ripley knew transitioning Brook Farm from transcendentalism to Fourierism would likely alienate some of the existing members. But by the time he'd decided to pursue that route, Ripley had convinced himself that Fourier was well-nigh a new messiah who had delineated a social model that would "cause the will of God to be done of [sic] earth, as it is in Heaven."[26]

Unlike the often abstract philosophy of transcendentalism, Fourierism is an elaborate, and often amusingly specific, system. It begins with the premises that humans have 810 discrete personality types, governed by twelve God-given "passional attractions," and that social evolution moves through thirty-two stages. Fortunately, by the 1800s, humans had already passed through quite a few of those stages; Fourierists hoped to hasten the transition from "Civilization" (a tawdry society with insular families, exploitative capitalism, and corrupt religion) to "Harmony" (a delight-filled, communitarian society in which everyone could pursue his or her own notion of freedom).

Although farming and food were at the core of his theory, Fourier rejected the family farm as an economic unit—not because it hadn't worked, but because it was inefficient. Even networks of small farms would soon be unable to feed the rapidly growing urban populations. However, Fourier was adamant that farms shouldn't be controlled by "monopolists." If they were, owners would "reduce all those below them to commercial vassalage, and achieve control over the whole of production by their combined intrigues. The small landowner would then be forced indirectly to dispose of his harvest in a way that met with the monopolists' agreement," worsening the exploitation and poverty of agricultural workers. He was right to worry.[27]

The middle way, Fourier's logical alternative, was to farm communally—as the Brook Farmers did. Ideally, each social/farm unit, which he called a Phalanx, would consist of 1,620 people (two each of the 810 personality types). Every member would have a complement and every task could be done by someone "passionally attracted" to it. No task would be done for more than two hours at a time, and every day would incorporate a mix of work and play. All could pursue their passions, but the primary focus of the Phalanx was growing, preparing, and eating delicious food. Fourier described his theory as a "gastrosophy," a gastronomic science, because he considered eating a fundamental pleasure, the foundation of society, and the heart of all politics. Food was the center of his good life; once humans reached Harmony, they would eat five meals a day. Therefore, people had to be experts in agriculture, food preparation, food preservation, and gastronomy.

Children were to begin cooking lessons as soon as they were able. First, they'd learn to prepare "their favourite sweet foods—things like sugared creams, compotes, cakes, jams and fruits." Once they were old enough to work with fire, they would tackle "the food most appropriate to their age and bodies, graduating . . . by ingredient size from larks to quails to pigeons, to chickens and then to larger joints as they themselves grew bigger." Much as every Utopian in Sir Thomas More's book was a farmer, every Fourierist was a gastronome.[28]

As had Sylvester Graham, Fourier identified failings in bread production as a failing of civilization. But instead of offering people a single version of unbolted flour–based bread, the Frenchman believed the Phalanxes should bake breads with "three degrees of SALTINESS, three of YEASTINESS, three of COOKEDNESS," accomplishing that with three different flours. The result would be "twenty-seven sorts of bread to give a group of three men a harmonic dinner." Whereas Graham was concerned with the healthfulness of the bread, Fourier worried about having loaves exactly suited to each citizen's taste.[29]

Fourier's gastrosophy and Alcott's dietary austerity are as different as every other aspect of the two worldviews—save two. Both considered a specific diet necessary to attain and maintain one's perfected state, and both suspected that industrial foods were usually adulterated. Fourier feared that if eaters were indifferent to how well or poorly foods were made, commercial makers would "produce their products in the cheapest way possible." He disdained the Parisian bakers who undercooked pastries so leftovers could be sold the following day to customers who didn't realize they weren't fresh. Fourier himself sought "the best quality of everything, free of adulteration."[30]

Though Brook Farm officially changed course from transcendentalism to Fourierism, it never reached 1,620 members, topping

out around 120. In 1845, the community suffered a smallpox out-
break that prompted parents to withdraw their children from the
farm's school, thereby reducing Brook Farm's income. Later, the
school lost most of its remaining students when parents heard
scandalous rumors—including that Fourierists believed in "love
germination," the innate inability of some people to be monog-
amous, and that they advocated having "Little Hordes" of children do jobs like mucking stalls and cleaning bathrooms. Being
steeped in dirty work early on, Fourier believed, would rid them
of filthy desires later. Finally, a devastating fire in 1846 drove the
struggling community to its demise.

I'm not surprised parents would object to their children being
required to perform those jobs (as its happens, the ones attending
the Brook Farm school weren't). What does surprise me is that
Ripley and the other Brook Farmers appear not to have had their
faith in Fourierism tempered by its many quirky elements. I've
yet to see any reference to a Brook Farmer finding it odd that a
three-year-old child spit-roasting larks was specified as part of the
philosophical grand plan. Or thinking it strange Fourier predicted
that once humans attained Harmony, we'd live to the ripe old age
of 144, would reach an average height of seven feet, and would
have strong tails. Or that when the global population reached "3
thousand million, there [would] normally be 37 million poets equal
to Homer, 37 million geometricians equal to Newton, 37 million
dramatists equal to Molière, and so on . . . (These estimates are
approximate)." Or that when we achieved this ideal state, a crown
of light would warm the North Pole and melt the ice caps and this
"boreal liquid" would flood the oceans and make them taste like
lemonade. Or that while warfare would continue, it would take
the form of massive cooking contests, with vast armies sending
"their daily offerings to the grand gastronomic jury."[31]

Though it never attained lemonade-ocean nirvana, Brook
Farm temporarily achieved its aims. Members were housed and
fed (though as the community's financial situation worsened, so,

too, did its diet). The ill and elderly were cared for. One early resident, George Partridge Bradford, wrote to a friend, "The whole establishment is conducted far more easily and more in accordance with the professed or rather implied principles on which it is based than will readily be conceived by those who have not witnessed its operation. We all meet on terms of perfect equality at the table." Like some other residents, he surprised himself by discovering that some "unsuitable" tasks could be quite pleasurable, adding, "You will laugh when I tell you that I am learning to milk."[32]

Another resident, Georgiana Bruce Kirby, recalled that visitors would seldom "conceal their amusement at the fanaticism exhibited by well-bred women scrubbing floors, and scraping plates, and by scholars and gentlemen hoeing potatoes and cleaning out stables; and particularly at the general air of cheerful engrossment apparent throughout." Their gaze, she wrote, made her feel "as if we were all acting in a play," which, in a sense, they were. Like Thoreau and the Fruitlanders, the Brook Farmers were performing utopia for the curious. And a lot of people were curious: One record listed more than four thousand visits in a single year, more than any other utopian experiment of the time.[33]

Its popularity isn't surprising. After Brook Farmers finished work, their evenings were filled with singing, dancing, and lingering in the fields gazing at the stars. Residents staged plays and performed classical music for one another. Emerson, Thoreau, Alcott, Fuller, and Brownson all came to give lectures. Marianne Dwight, a Brook Farmer, wrote to a friend that she "would not exchange this life for any I have ever led. I could not feel contented again with the life of isolated houses and the conventions of civilization . . . life is so full and rich here, that I feel as if my experience were valuable, and I were *growing* somewhat faster than when I lived in Boston." Like Marianne, most other participants were unaccustomed to intermingling between genders and across class lines. Many found the social openness intoxicating—so intoxicating that more than a dozen Brook Farm couples married during the five years the community was active. After it ended, many remained in the area or moved to another vibrant community nearby—in the town of Concord.[34]

TO LIVE FOR ONE'S PRINCIPLES

BROOK FARM WAS THE first secular utopian experiment in the United States, and was joined in the 1840s by just a handful of others. Comparing the secular experiments to the religious ones of that era, the utopian scholar Lyman Tower Sargent found several contrasting patterns. The religious utopias were usually organized hierarchically and were authoritarian, though the experience of living in one was usually one of economic equality. Participants in religious utopias rarely engaged in social-reform efforts. In contrast, the secular experiments were governed more democratically, though the sense of lived equality was sometimes undercut by practices like offering different meal options at different prices, as happened at Brook Farm. Members of secular utopias usually participated in social-reform movements. Chief among those concerns were improving education, abolishing slavery, and advancing women's rights. Both the Fruitlanders and the Brook Farmers were active social reformers, and many of them used their writing or speaking skills—their platforms—to promote abolition.[35]

George Ripley and Charles A. Dana, another original member of Brook Farm, used the *Harbinger*—the newspaper they edited and published there—to advocate for broad social change. A lengthy *Harbinger* article called for the end of slavery not only in the United States, but also internationally: The author urged "the abolition of universal slavery,—white slavery and black slavery, chattel slavery and slavery of capital, slavery of the soul and slavery of the body, and the slavery of the soil in addition."

Acknowledging this might seem too vast to be achievable, the author scaled back to the United States, and urged abolitionists to recognize "the two kinds of slavery the most prevalent in this country,—chattel slavery at the south, and the slavery of capital, or the wages system at the north." After Brook Farm failed, Dana went on to become an editor of the *New York Tribune*, giving it a strong abolitionist voice.[36]

Bronson Alcott was inspired by his wife to take direct action to protect slaves and promote abolition. Abigail (May) Alcott grew up in a household opposed to slavery, and Bronson came to embrace her view with equal vigor. Described as "an antislavery pioneer," Abigail was a member of not only Concord's Female Anti-Slavery Society, but also those of Philadelphia and Boston. Like many other Concord women, she raised money for abolitionist groups and efforts, signed petitions, attended antislavery conventions, and—on at least two occasions—provided safe harbor for runaway slaves. In 1832, Bronson Alcott, Samuel May (Abigail's brother), and others joined William Lloyd Garrison in founding the New England Anti-Slavery Society.[37]

In 1851, when the Fugitive Slave Law was applied against Thomas Sims, Bronson joined the society's Vigilance Committee, walking the streets at night to protect Black Bostonians from being seized and charged as fugitive slaves and to make sure actual fugitive slaves were brought to a safe house. In 1854, when Anthony Burns was jailed as a fugitive slave, Alcott and another member of the Vigilance Committee tried to rescue him. And in 1859, after John Brown was killed at Harpers Ferry, his widow and daughters moved to Concord to live with the Emersons for a time and then boarded at the Alcotts' for several months. In a letter to her brother from that period, Abigail wrote, "I dread a war—but is not a Peace based on such false compromises and compacts much more disastrous to the real prospects of the Country generally—and Freedom in particular—I think so."[38]

Ripley's, Dana's, and the Alcotts' behavior, both as utopians and as abolitionists, reveals their faith in their own capacity to help bring into being an alternative to existing conditions. That sensibility was widely held in their era, as both the religious and the secular communities believed they could start anew, creating heavens on earth. The groundswell of urgency had several sources, including growing opposition to slavery, women's increasing involvement in fighting for political causes (including their own rights), the millennialist emphasis of the Second Great Awakening, and an influx of fresh ideas brought by immigrating European socialists and utopians.

Another less obviously ideological issue also contributed. The transition from subsistence agriculture to agricultural capitalism tipped several regions into the economic paradigm of wage labor. In remote areas, such as Maine, Vermont, and much of the Midwest, most people still lived on farms. But in eastern Massachusetts, Rhode Island, southern New York, the eastern part of Pennsylvania, Maryland, and the District of Columbia, people were congregating in cities. In 1840, fifteen of the nation's twenty largest cities were in that four-hundred-fifty-mile-long swath; there, for the first time in American history, people were more likely to work in manufacturing, trades, or commerce than on the land. The BosWash Corridor was born.

Working as wage laborers left many people feeling less connected to the natural world than they had been and more economically precarious. Moreover, urban poverty worsened considerably during the six-year recession following the financial Panic of 1837. Against such a backdrop, secular utopias and religious intentional communities must have seemed incredibly appealing. When life was hellish, the promise of heaven on earth would be almost irresistible.

FOOD RULES IN HEAVEN

MOST OF THE ANTEBELLUM utopians didn't need to establish food rules. The religiously based communities had rules derived from their sacred texts or given by their faith leaders. The immigrant utopian communities tended to maintain their cultures' food habits. But for the Fruitlanders and Brook Farmers, secular communards most of whom were born in the United States, developing food rules was important. They—along with Thoreau—derived inspiration from transcendentalism, but the differences among their rules make it clear that the new philosophy didn't provide either agricultural or culinary best practices.

Because transcendentalism mostly opined on the ideal relationship of humans to the natural world, to their own divinity, and to one another, it was up to the utopians themselves to figure out what that looked like with respect to food. The importance of self-reliance is easy to see in their desire to raise as much of their food as possible. On the other hand, though they all valued a sense of equality among all beings, their dietary decisions don't reflect a consistent method for enacting that ideal. The Fruitlanders ate a subsistence vegetarian diet, adhering to it so firmly that they banished one member because she ate a piece of fish at a friend's house. They didn't exploit animals and slaves, but their principles came at a high price: The underfed Alcott girls were often ill, and Abigail Alcott was utterly exploited.

The Brook Farmers raised livestock as well as vegetables and ate meat when their finances permitted, reserving just one dining table for "Grahamites." Meals were communal and, by most accounts, quite jovial, and the community integrated food into their celebrations. While men and women were equal in terms of pay scale and were supposed to be equal in terms of hours worked, nine out of the ten people logging the most hours working at Brook Farm were women—and the one man on that list was the baker. As was the case for Abigail Alcott, the women at Brook Farm worked more, not less, than they had before arriving there.[39]

Both farmer and cook, Thoreau raised vegetables and legumes, gathered fruit, and sometimes fished. He ate simply at Walden Pond, but one of his staples, rice, was a product of slave labor. When dining with family or friends, he ate whatever he was served, which no doubt included other items Alcott banned at Fruitlands.

To decide if one of these approaches was most consonant with transcendental ideals would involve parsing what they ate, how those foods were produced, how the field and kitchen workers were treated, how the meals themselves were organized, and how the food's production and consumption aligned with their philosophical aims. Attempting this isn't straightforward, for each approach has strengths and weaknesses. Does the important social role of meals at Brook Farm make up for overworking the women? Does not exploiting animals compensate for seriously under-nourishing your children? Does eating rice undo the many non-exploitative choices Thoreau made? In the imperfect matches these idealists made between diet and values, we see hints of the food conflicts that have proliferated since. As American agriculture and American food preferences became more diverse, deciding on the "right" way to eat got more difficult.

Though these utopian back-to-the-landers were imperfect models, their ideals—to live lightly, to eat justly, to regard others as equals, and to cultivate freedom—have continued to animate utopian efforts. They are rekindled, again and again, when social conditions shift and finding alternatives feels especially urgent. By 1900, economic inequality in American cities had grown considerably worse, and many reformers urged urbanites to turn to the land to establish self-sufficient lives. Some became homesteaders, and others lived in communes and collectives.

Among the most intriguing of the back-to-the-land experiments of that era was Arden, Delaware, a village whose founders were inspired by transcendentalism, by Henry George's innovative economic theory, and by the British Arts and Crafts movement. Incredibly, Arden remains a vibrant community today.

4

GOING BACK TO THE LAND DURING AMERICA'S FIRST GILDED AGE

AFTER THE 1840S, THE number of utopian back-to-the-land experiments waned for several decades. Then, between 1890 and 1910, with the nation firmly on a path away from its rural, agricultural origins and toward its urban, industrialized future, another groundswell of utopian experiments occurred. As in the 1840s, many of them were religious or spiritual in nature; in this second surge, the Christian experimenters were complemented by Jewish, theosophist, and other groups. The secular utopian efforts were also more diverse, with anarchist, ethnic, economic, and aesthetic concerns shaping them. The two most common sorts of secular experiments were single-tax communities and art colonies.

One colony, the village of Arden, Delaware, combined the single-tax philosophy with ideals gleaned from the British Arts and Crafts movement, and added a healthy dollop of transcendental thought. In 1900, the sculptor Frank Stephens and the architect William Lightfoot Price purchased a 162-acre farm just outside Wilmington to create their utopia. They named it Arden after the forest in Shakespeare's *As You Like It*, an anarchic space where characters are temporarily freed from social mores. At Arden's entrance stile was a graceful, arched placard inscribed with other Shakespearean references: "YOU ARE WELCOME HITHER" greeted arriving visitors and "IF WE DO MEET AGAIN, WHY WE SHALL SMILE" bade them farewell.[1]

Stephens and Price succeeded far better than have the founders of most utopian experiments; the vibrant, quirky, progressive community they envisioned retains some of its original character to this day. Like Thoreau and other transcendentalists before them, Stephens and Price recognized that rapidly changing social conditions were making most urban Americans feel disconnected from the natural world and from their own lives and labor. The two men believed that embracing the single-tax model promoted by Henry George and integrating art and nature into everyday life, as was being espoused by the British Arts and Crafts movement, would enable people to feel more connected to themselves, their surroundings, and one another. In that way, people would have a better chance to enjoy an economically just, self-sufficient, collaborative, and beautiful life.

RISING ECONOMIC INEQUALITY AND HENRY GEORGE'S SINGLE TAX

IN THE SECOND HALF of the nineteenth century, only the richest urbanites could have accurately described their lives and surroundings as "beautiful," and only the most obtuse would have called city life "economically just" or "self-sufficient." Corporations and financial institutions were expanding so rapidly that they assumed the economic and political power formerly held by farmers and small-business owners. Rather than reining in those big companies, the government took its cue from them. As the economics commentator Kevin Phillips has noted, the government doled out enormous favors "that helped produce the world's biggest fortunes—by all but seizing key portions of federal and state government." With his characteristic cutting brevity, the historian Arthur Schlesinger Sr. wrote: "America, in an ironical perversion of Lincoln's words at Gettysburg, had become a government of the corporations, by the corporations, and for the corporations."[2]

Among those who especially benefited from governmental giveaways were railroad owners. Not only did the federal government pay them to lay track, it also gave them twenty square miles of land for each linear mile laid. In all, the government gave the railroads more than 155 million acres of land—more than are in Alaska. Municipal governments poured still more into the corporate coffers as they tried to persuade railroads to come through their towns. Thanks to all this largesse, the rich grew richer still: The wealthiest 1 percent of American households amassed spectacular fortunes and controlled 51 percent of the household wealth. Folks like J. P. Morgan, John D. Rockefeller, Andrew Carnegie, and William K. Vanderbilt had so much money, and flaunted it so ostentatiously, that Mark Twain dubbed the era "the Gilded Age."

For the other 99 percent of urbanites, city living was anything but lavish. About a third lived in dire poverty, perpetually on the verge of starvation. Another third struggled, particularly during the several economic downturns of that era, most notably the one from 1873 to 1877, another from 1884 to 1886, and another from 1893 to 1897. During each, urban workers were laid off or fired. The remaining third, primarily artisans, entrepreneurs, and white-collar workers, attained a middle-class level of income and comfort, though their status was also precarious. As the economist Abraham Epstein observed, "Any one of the accidents to which they are liable, often throws such people, within a week or two, on charity." Among the many challenges urbanites faced was simply getting enough to eat: Between 1890 and 1906, the percent of its income a "workingman's family" spent on food rose from about 30 percent to more than 42 percent. Few wage earners could attain the American Standard of Living.[3]

Even so, most economic thinkers of that era insisted the growing economy would lift rich and poor alike. One self-taught economist, Henry George, realized this wasn't the case. He pointed out that increases in wealth were always accompanied by increases in poverty: "The 'tramp' comes with the locomotive, and almshouses and prisons are as surely the marks of 'material progress' as are costly dwellings, rich warehouses, and magnificent churches." He set out to find why, and concluded that advancing wealth *caused* advancing poverty. As the population grew, so, too, did the value of land; land speculation enabled the rich to get richer and, at the same time, caused the poor to become relatively poorer. But, as George argued in his book *Progress and Poverty*, if the US tax code stopped rewarding speculation, then the wealth gap could be reduced.[4]

He proposed instituting a hefty tax on land and eliminating all other taxes. If landowners paid a high tax, speculating might become too expensive to be worthwhile. And instead of being hoarded, land might be used for its best purpose—like a commons. George sensed that rising industrialization was making the situation more urgent. Unless the poor and the middle class had some economic protection, laborsaving devices and the new efficiencies in manufacturing would operate like "an immense wedge" forced "through society. Those who are above the point of separation are elevated, but those who are below are crushed down." If the system didn't change, the gap would continue to widen, making "sharper the contrast between the House of Have and the House of Want."[5]

Will Price, Frank Stephens, and their benefactor, the zealous single-taxer Joseph Fels, were hardly the only ones captivated by Henry George's ideas and interested in putting them to the test. *Progress and Poverty* was an international bestseller, translated into a dozen languages. In the 1890s, it sold more copies than any other book except the Bible. George's ideas inspired not just Arden, but also the single-tax communities of Fairhope, in Alabama; Halidon, in Maine; Gilpin Point, in Maryland; Tahanto, in Massachusetts; and Free Acres, in New Jersey. Although not all of these communities were back-to-the-land experiments, Free Acres and Fairhope, like Arden, were farming- and gardening-centric.

To employ Henry George's model in Arden, the founders created a trust to own the land. Residents leased their lots for ninety-nine years and paid their rent to the trust, and that money paid for community services. Decisions about how to allocate it were discussed and voted on by all members of the community (even, at the outset, the children). On their rented lots, residents could build as they wanted, but the lots themselves remained the property of the trust.

Most of Arden's early residents were single-taxers, socialists, communists, anarchists, and other radicals who found this model appealing, or at least more appealing than capitalism. In those first few years, when Arden was a summer-only community, reformers used their stay as an opportunity to test their living principles with similarly-minded folk before returning, each fall, to the demands of the wider world. As one early resident wrote: "The country . . . promises health, the grandeur of nature, the sublimities of rest and tranquility. But it is the peace and the silence of the armistice. The battle awaits. In the city is the multitude, the multitude who knows no rest from the exacting labor who know[s] no tranquility."[6]

Though not of one mind politically, the residents of Arden all sought to have and to model a better way of life than that available in the mainstream: "We, who have left behind so much of what men call the comforts of civilization, have left behind something of its fear and its despair . . . We have found by wood and open field, in that cottage and cabin, something of that which it has lost, something of hope."[7]

ARTFUL ARDEN

Arden's growth and character were informed by Stephens's and Price's great interest in the Arts and Crafts movement, especially as it played out in Britain. Like other Americans who were Arts and Crafts proponents, Price and Stephens melded the British ideas with homegrown transcendental thinking. The amalgam worked well because the British Arts and Crafts movement borrowed a lot of ideas from Walt Whitman and Ralph Waldo Emerson. In addition, Emerson's thinking had been heavily influenced by that of Thomas Carlyle (Emerson's friend who'd found Bronson Alcott so odd).

Thomas Carlyle was a polymath—a historian, an essayist, and a philosopher—who lived in England for much of the nineteenth century. By virtue of that, he witnessed what industrialism did to people and communities before Americans did. As factories proliferated, Carlyle described their products pejoratively, calling them "cheap and nasty" and "Devil's-dust cunningly varnished over." His problems with factories were twofold. First, the system hurt laborers physically: Factory work conditions were awful, with long days, unhealthful workspaces, and frequent corporal punishment. Second, the system hurt laborers psychically or spiritually. Carlyle believed doing work one was proud of was the key to the good life; factory workers were deprived of that opportunity, leaving them alienated from their own labor and thus also from their own lives.

Carlyle's attitudes notwithstanding, factories proliferated in England, and people looking for employment moved from the countryside into cities ill-prepared to house them. Whole families often lived in a single, poorly lit, poorly ventilated room without indoor plumbing. The lack of clean water and the presence of excrement and other garbage in the streets contributed to disease outbreaks and epidemics, and pollution from burning coal enshrouded the cities in miasmal gloom.

Influenced both by these deplorable conditions and by Carlyle, the English thinker John Ruskin wanted to foster a more benevolent social order. As had many utopians in the 1840s, Ruskin advocated socialism; he proposed developing a cooperative society that emphasized art, craft, and sustainable farming. There, people could reintegrate mental and physical labor—much as the Brook Farmers had hoped to do. He believed workers should be in charge of their own time and tasks, as he imagined had been the case during medieval times. To achieve that, he wanted to organize the community as a set of guilds in which apprentices could learn blacksmithing, carpentry, textile production, and other crafts. Ruskin's hoped-for community never fully materialized, but his ideas inspired several protégés—most notably, William Morris—to create their own guilds

Like Ruskin, Morris imagined the medieval period had been a time when work was pleasurable and craftsmen could follow their own aesthetic impulses. This rosy image of medieval labor as unalienated somewhat missed the mark, but their idea about the importance of having pleasurable work resonated with the thinking of many other socialists, including Fourier and Engels. Morris shared the transcendental emphasis on nature as a spiritual touchstone, and believed "art made by the people, and for the people" would ensure "happiness to the maker and the user." If people didn't reconnect beauty and labor, he feared "the result of the thousands of years of man's effort on the earth must be general unhappiness and universal degradation."[8]

Unlike Ruskin, Morris did succeed in creating several guild-like companies. The first was Morris, Marshall, Faulkner & Co., an interior-design group better known as "the Firm." The artisans worked in Red House, which was built specifically to serve as both the design studio and Morris's home. In the spirit of a medieval guild, the Firm's designers created beautiful, handmade carvings, frescoes, furniture, stained glass, and textiles. Key to their vision was the idea of art as integrated into life, as *useful*, rather than art as a commodity to be owned and displayed as a marker of status.

Frank Stephens was introduced to the Arts and Crafts ideas of Morris and Ruskin during his time as a student at the Pennsylvania Academy, in Philadelphia, where he studied from 1879 to 1885. Captivated by Morris's ideals and by what he created, Stephens and three fellow alumni established a decorative-arts business akin to the Firm. Other admirers carried both the Arts and Crafts aesthetic and the guild model to the United States, where it splintered into discrete configurations, some more philosophical, others more aesthetic, some centering on consumers, others on producers. The version infusing Arden emphasized being a producer, working hard, and cultivating nature-oriented beauty.

Will Price first encountered the Arts and Crafts movement in 1896, on a trip to England. There, he met John Ruskin and William Morris and was immediately intrigued with their ideas. For years after his trip, Price would read aloud "with great gusto from Morris' writings after Sunday dinner." Likewise, he became captivated by the architectural language he saw in English villages, where the houses often used similar, local materials and were less fussy than the elaborate mansions he'd been building for wealthy East Coasters.[9]

Both Price and Stephens took the philosophy behind the Arts and Crafts movement quite seriously, and wanted Arden to be a preindustrial, creative culture with egalitarian and communal values: "We had learned William Morris' truth that nothing can be done for Art until we have bridged the terrible gulf between rich and poor," Stephens wrote. "We were so disgusted with civilization that we determined then and there to go out into the open and make a better one." That "better one," he believed, would provide "easier ways of earning a living, a simpler, more democratic and more peaceful manner of life than that characteristic of our time and country, and a freedom from mere wealth-slavery from which craftsmanship and art will of themselves develop."[10]

Stephens and Price encouraged residents to establish "gilds" (their preferred spelling) to shape the community's social life and help enact their philosophy. Among the early ones were an Athletic Gild, a Craftsmen's Gild, a Musicians' Gild, and a Scholars' Gild. As they'd hoped, the community became a bastion of creativity. Shakespearean plays were produced at least once a year; in fact, the residents constructed an outdoor stage before they constructed any buildings. The town created its own holidays, including a celebration day for Walt Whitman and one for Henry George; held the Arden Fair and the Arden Field Day in the summer; and had annual pageants, all of which were open to the public. Like the open, convivial atmosphere at Brook Farm, Arden's led to plenty of romances, prompting one wag to say, "Free love was not part of the single-tax program, but it might as well have been."[11]

The creative energy of Arden was not limited to the arts; the residents also advocated for equal rights for women, better child-labor laws, and other social reforms, as did Stephens and Price. Both men were pacifists and war objectors. Frank Stephens so emphatically opposed President Wilson's decision for the United States to enter World War I that he wrote to him, saying, "I will neither kill nor help kill." When he refused to buy Liberty Bonds, he was arrested for "attempt[ing] to interfere with the prosecution of the war," though he was ultimately found not guilty. Stephens's willingness to go to jail for his beliefs and his repudiation of the government when it acted dishonorably resonated with similar actions of Thoreau and Alcott before him, as well as with those of subsequent back-to-the-land utopians, including Scott Nearing, who also objected to the United States entering WWI, and many hippies in the seventies who avoided serving in the Vietnam War because they did not condone it.[12]

Thanks to Will Price, Arden had an architectural master plan. He envisioned the village growing around central green spaces, like an old-fashioned town common. The house lots were small, all an acre or less, and curving pathways connected the town center to the woods on its edges. By emphasizing public spaces and de-emphasizing private spaces, the layout encouraged socializing and working collaboratively. Residents soon traded their tents for cottages they constructed together, often incorporating references to medieval English architecture. Unlike the residents of Fruitlands and Brook Farm, many of those in Arden had practical skills, thanks to two features of Philadelphia's culture: The city maintained small-scale and specialty production longer than most US cities did, thus still had tradespeople who'd apprenticed in local mills and workshops. And even the children of elite families were expected to learn a trade before embarking on a profession; an aspiring engineer might learn the skills of a machinist, for example, while a budding architect would likely learn carpentry.[13]

One of Price's innovations was a design for "the house of the Democrat," and he gave the specs to anyone who wanted them. Like Thoreau's cabin, the house of the Democrat was simple, built of local materials, and designed not to "oppress by its possession." Local fieldstone was used for the foundation and walls, supplemented by oak and poplar harvested from Sherwood Forest or Arden Woods. The local materials made the buildings feel more integrated into the landscape, while also being, as Price pointed out, inexpensive, easily obtainable, and beautiful. Even as the buildings grew more polished, with bits of colored tile added to plastered walls for decoration, the domestic spaces stayed small and the communal spaces large and welcoming.[14]

Price considered this way of organizing personal and communal spaces, and this emphasis on personal creativity, necessary for freedom. Describing the house of the Democrat, he argued having an affordable home that you aren't afraid of losing is critical to being able to live well and generously: "Out of the free gift of their surplus hours," safely housed people can and will "build for each and for all, such parks and pleasure places, such palaces of the people, such playhouses, such temples, as men have not yet known." Moreover, "the men and women and children shall find playtime to use them; find time and powers out of their work to write plays and play them, to write poems and sing them, to carve, to paint, to teach, to prophesy new philosophies and new sciences; to make, to give, to live." Like Thoreau, Price believed people didn't need to work incessantly—if they pared down their wants, they could become producers who had ample time for truly living, for reclaiming freedom.[15]

As Arden grew, other evidence of transcendental influences became apparent. The path traversing the village green was named Walt Whitman Way. In a similar nod to William Morris, Stephens's studio, the forge, and some communal spaces were in a building christened Red House; when it needed to be replaced, its successor was dubbed Gild Hall. This was soon joined by the Craftshop, which functioned as a guild space—housing a woodworking shop, a sewing room, a weaving room, a metalworks room, artist studios, and a bakery. A decade into the new century, Arden's broad village green was abutted by not only a gaggle of small homes and cottages, but also an ice-cream parlor, an outdoor theater, a hotel, a community center, shops, a tennis court, and a baseball diamond.

EATING IN ARDEN

As did Thoreau, the Fruitlanders, and the Brook Farmers, the Ardenites established food rules consistent with their overall beliefs. Frank Stephens was a vegetarian, and he expected agriculture to be a shared and central enterprise: "Everyone is to some extent a tiller of the soil," he noted, "as it was in that picturesque medieval life which came from an underlying communal land tenure on which our system is an improvement." The emphasis on communal land tenure ties in to his commitment to the single-tax model, and the notion that tilling the soil is picturesque and medieval is drawn from the Arts and Crafts ideals. Some of the early residents were quite adept growers—many "rented small plots and raised truck to be hauled into the markets of Wilmington, six miles away." One of those truck gardeners, Hal Ware, was just ten years old when he began selling the vegetables and flowers he grew. Others grew primarily for their own sustenance; one couple set out "to ascertain whether it were possible for two people to grow enough produce on their acre of land to enable them to live without other assistance," and discovered that they absolutely could.[16]

The Ardenites' self-reliant food production was complemented by collective food preparation. Meals were often communal, providing yet another opportunity to socialize. The houses in Arden in its early years did not have kitchens; residents could prepare meals in one of the communal kitchens or eat at the Arden Inn, where the mostly vegetarian meals were served family style. During work parties and festivities, food was also prepared collectively, and enjoying food together was part of the event. Although Arden's residents didn't exit the mainstream economy, their emphasis on growing so much of their own food and vegetable-centric menus enabled them to be less complicit in the race-based injustices permeating the nation's agricultural systems during the latter half of the nineteenth century.

AMERICAN EXPANSION AND AGRICULTURAL EXPLOITATION

SLAVERY ENDED AFTER THE Civil War, but for many agricultural workers conditions did not improve significantly. Exploitation of Black farm laborers continued in the South, and racist labor practices and the genocide of Native Americans became ubiquitous in the Midwest and West. Thanks to the railroads and to the Homestead Act of 1862, both underwritten by the US government, droves of both internal migrants and immigrants headed to the Midwest and the West in search of opportunity and arable land.

Much of that territory, however, was already occupied by Native peoples who didn't want their spiritual homes and hunting grounds turned into 160-acre lots for white settlers. In skirmish after skirmish, Indigenous people and settlers killed one another. Even more Native Americans were killed during the 1870s when the United States waged a series of wars against their tribes. Those who survived were forced off their land and relocated to reservations far from their homes, places where their sacred foods often could not thrive. As westward expansion by whites continued, the reservations were deemed larger than necessary, and three-quarters of the reserved land was taken back so it could be given to white people. During the lifespan of the Homestead Act, more than 270 million acres—about 10 percent of US land—was stolen from Native peoples and given or sold to white settlers.

On grassy rangelands where Native peoples had grazed and hunted bison, white ranchers wanted to graze cattle. The military and individual ranchers killed many of the people living there so they could create ranches. By 1890, four large meatpacking companies in Chicago controlled the processing and transporting of cattle and meat via the rapidly expanding railroad network. The ranches were too small to negotiate well on their own and too far from one another to form alliances or blocs, so the meatpacking plants, like the railroads on which their success depended, often pursued exploitative practices.

California's agricultural economy initially relied on Native Americans. In the 1850s, wheat farming offered a particularly reliable source of income, in part due to a law passed in the state's first legislative session called the "1850 Act for the Government and Protection of Indians." The "protection" was hardly such; the law gave white farm owners access to free labor, and allowed them "to procure unfree Indian workers through a system of lo-

cal government convict leasing or to ensnare free Indian laborers contractually in what amounted to legalized debt peonage."
Supplementing that legal system of de facto slavery was an illegal
slave trade of "Indian wards and apprentices, who were sought
primarily as household domestics or as farm workers." Forced labor by Native Americans enabled California's agricultural sector
to expand rapidly between 1850 and 1870.[17]

However, conflicts between white settlers and Native peoples
continued. In his State of the State Address in 1851, Governor
Peter Burnett pledged that "a war of extermination will continue
to be waged between the races until the Indian race becomes extinct." By 1873, the Native population had plummeted from about
150,000 to 30,000. Most of the decline came from state-sponsored
killings. With Native populations all but decimated, farm owners
turned to other ethnic groups in search of a reliable labor pool,
and during the 1870s, many of the Chinese immigrants who had
helped build the railroad became field hands.[18]

Between 1870 and 1900, wheat farms grew immensely, their
yield more than doubling. The agricultural historian R. Douglas Hurt described them as "bonanza enterprises characterized
by absentee ownership, financing based on extensive credit, and
mechanization, particularly the combine"—the beginnings of
the industrial agricultural model that persists today. Once irrigation became readily available, land prices skyrocketed from an
already high seventy dollars per acre to fifteen hundred dollars
an acre. With entry costs beyond what individuals could afford,
California agriculture was dominated by large, capital-intensive
enterprises.[19]

In the American South, a toxic combination of laws and prej-
udices contributed to a new agricultural model—sharecropping.
During the Civil War, the Union Army had seized four hundred
thousand acres of southern land, much of it farmland. As the war
was ending, General William T. Sherman and Secretary of War
Edwin Stanton met with twenty African American ministers, led

by Reverend Garrison Frazier, to find out what the newly freed African Americans most wanted. Learning from the ministers that land and self-governance were their main desires, General Sherman issued "Special Field Order #15," which stipulated that each emancipated family be given forty acres of tillable land from the seized acreage. In those new communities, "no white person whatever, unless military officers and soldiers detailed for duty, will be allowed to reside; and the sole and exclusive management of affairs will be left to the freed people themselves." Lincoln promptly signed Sherman's field order, and it became—briefly— law. Between its signing in January and the following June, forty thousand freedmen settled on the land.[20]

Scant months after Lincoln's assassination, President Andrew Johnson overturned the order. Some of the land was returned to its former white owners and some was auctioned off and pur-chased by speculators, often from the North. Without land of their own, southern Blacks who'd only ever labored in fields had little recourse but to work as sharecroppers or tenant farmers. Laborers were perpetually in debt to the landowner and obliged to plant whatever the owner wanted. The majority wanted cot-ton, which stored and sold well. But overproduction led to signifi-cant price drops, driving sharecroppers deeper into debt. Because fields planted in cotton couldn't be eaten, sharecroppers had to tend their own small food gardens in the evening after working for others all day—if they were allowed to have such gardens. Some owners demanded that cotton be grown all the way up to the cabins where sharecroppers slept.

Some freed Blacks could afford to buy land and did so. Al-though the end of the nineteenth century was tumultuous and of-ten violent, as many white landowners and civil servants sought to undercut advances by emancipated Blacks, the number of Black-owned agricultural enterprises grew. In 1900, 746,717 Black Americans were farmers and by 1920 that number had risen to 925,710—the highest it's ever been.

During that era, all small farms faced serious challenges. Over-production drove down the price the farmers could get for goods, and an increasingly national market meant dealing with more middlemen. Railroads, stockyards, grain elevators, and assort-ed distributors all took a cut of every food dollar. White farmers, however, could usually count on help from the USDA. Since its establishment, in 1862, the USDA disproportionately helped white farmers. Non-white farm owners had fewer opportunities to get loans, educational assistance, or other aid from the USDA, so had fewer ways to expand, diversify, or otherwise shore up their farms.

Because the USDA played—and continues to play—such an important role in shaping America's farm culture, its policies, fi-nancial incentives, and enforcement decisions largely determine which kinds of farms and farmers thrive. The changes to agricul-ture in the second half of the eighteenth century—toward larger farms, more white absentee owners, greater farm consolidation and vertical integration, and more reliance on exploited labor—all aligned with the USDA's vision for American agriculture.

In the context of the commercial agriculture of their era, the efforts of Ardenites to grow their own food were not politically or ideologically neutral; they were challenges to the dominant industrial agricultural system. Like Thoreau foraging, gardening, and building his own house, like the Fruitlanders avoiding foods and goods produced with exploitative practices, the residents of Arden demonstrated that non-exploitative alternatives to main-stream agriculture and culture could exist.

FOOD SAFETY

THE ARDENITES' SELF-RELIANCE ALSO enabled them to avoid adulterated foods, which otherwise would have been difficult to do. The United States had few national laws requiring food sup-pliers to specify the content of their products, so makers often cut expensive ingredients with cheap fillers. Items sold as honey and maple syrup, for example, were often mostly corn syrup. Wheat flour was often corn flour treated with sulfuric acid to make it look lighter and then laced with white clay and crushed white rocks. Milk was watered down with warm water, then whitened with chalk; its cream layer could be simulated by adding a bit of pureed calf brains. More dangerous were processors' reliance on chemicals whose safety for human consumption was unknown. Some chemicals were added to make rotten items seem fresh. Margarine makers regularly used leftover bits from meatpacking plants, and because margarine became rancid quickly, they added disinfectants to cover up the smell of the decomposing flesh. Similarly, borax and formaldehyde were added to rancid meat to make it smell fresher.[21]

As people became more aware of these issues, they joined "a crusade in favor of pure food" and prompted various states to devise their own regulations. But Congress was extremely slow to act. In 1900, Senator William Mason, from Illinois, begged Congress to pass national food-labeling and safety laws, pointing out that the United States was "the only civilized country in the world that does not protect the consumer of food products against the adulterations of manufacturers." His bill failed to pass.[22]

Because Congress wasn't acting, Dr. Harvey Wiley, head of the Department of Chemistry (precursor to the Food and Drug Administration), decided his office would figure out which—and how much—of the most common food additives and preservatives were safe. In 1902, he created a "poison squad." A dozen young men, all civil servants, volunteered to serve as human guinea pigs in exchange for three poisoned meals a day. Over the course of each chemical trial, the men ate larger and larger doses to determine how much was safe. The squad started with borax, then moved on to saltpeter, sulfuric acid, formaldehyde, copper sulfate, benzoic acid, saccharin, and salicylic acid. Although the group always numbered twelve, it wasn't always the same dozen.

The men could leave after a year and often did, sicker and weaker than when they'd begun.

When the experiments started, Dr. Wiley thought ingesting small amounts of the various additives was probably safe, but that at some tipping point, the dosage would be too much and would be dangerous. But during the borax trial, half of the men dropped out before the final round because they were already ill. Wiley realized that some chemicals accumulate in the body over time, building up to dangerously high levels. He urged that chemical preservatives be used only when absolutely necessary and that they be listed on the label so consumers would know. But Congress still wouldn't act. In fact, between 1860 and 1906, some two hundred bills advocating greater food and drug safety were presented to Congress. Just eight passed.

Then, in 1906, the effort was unexpectedly aided by an Arden resident, the labor organizer Ella Reeve Bloor, and a soon-to-be-Arden-resident, the muckraking journalist Upton Sinclair. That year, Sinclair's novel *The Jungle* was published. In it, he described conditions in Chicago's meatpacking plants in disgusting detail. His aim was to show that the plants' owners cared only about profits and not about workers or customers, but his socialist message was mostly ignored by grossed-out readers. Among the many horrors, his descriptions of sausage-making were especially revolting: "there would come all the way back from Europe old sausage that had been rejected, and that was moldy and white—it would be dosed with borax and glycerine, and dumped into the hoppers, and made over again for home consumption," he wrote, adding:

> Meat that had tumbled out on the floor, in the dirt
> and sawdust, . . . meat stored in great piles in rooms;
> and the water from leaky roofs would drip over it,
> and thousands of rats would race about on it . . . the
> packers would put poisoned bread out for them; they
> would die, and then rats, bread, and meat would go

into the hoppers together. . . . There were the butt-ends
of smoked meat, and the scraps of corned beef, and
all the odds and ends of the waste of the plants, that
would be dumped into old barrels in the cellar and left
there . . . and in the barrels would be dirt and rust and
old nails and stale water—and cartload after cartload
of it would be taken up and dumped into the hoppers
with fresh meat and sent out to the public's breakfast.

Riveted and aghast, readers demanded change.[23]

Enraged readers sent copies of *The Jungle* to President Theodore
Roosevelt, calling on him to do something. He ordered the USDA
to investigate. Sinclair warned him some USDA inspectors were un-
trustworthy and proposed to the president that he send undercover
investigators, with Ella Reeve Bloor as liaison between informants
and agents. When the investigators' report corroborated Sinclair's
account, Roosevelt told Congress to pass *something*. Even with the
president pressing them to act, Congress couldn't. Frustrated and
impatient, Sinclair leaked what he knew about the investigators'
findings to the *New York Times*, which forced Roosevelt's hand.

Americans were revolted by what they learned—but Euro-
peans were livid. Great Britain stopped importing American
canned meats; Germany and France stopped importing all meat.
Just three weeks later, after years of inaction, Congress passed,
and Roosevelt signed into law, the Meat Inspection Act and the
Pure Food and Drug Act.

THE EVOLUTION OF ARDEN

ONCE ARDEN WAS UP and running, Will Price moved on, eager to design and build another community. Frank Stephens
remained—he hoped to promote the single tax by promoting
Arden. Like other utopians, he understood the importance of

outsiders seeing utopia in action, so all the community's arts events and festivals were widely advertised and open to the public. As so many had in the 1840s, both the curious and the committed came to visit.

In Arden's early years, the Arts and Crafts ethos was primarily about living in a preindustrial way, being producers, learning from one another, and cultivating strong ties to the natural world. Few artisans were making goods to sell. But as the community grew, more people looked for ways to work from home, and cottage industries emerged. In 1912, the community built a forge and people began working with iron. Arden opened a craft shop, and finely made crafts became a source of income. The rising role of artistry led more artists to move in.

By 1922, a hundred people lived in Arden year-round and an additional three hundred and fifty came for summers. When Arden reached the capacity of the original property, Donald Stephens (Frank's son) established Ardentown on a 128-acre parcel abutting Arden to the southeast. The Depression hit the Ardens hard: Many of the artisans attempted to earn a living selling crafts, but with no one to buy them, the Craft Shop closed. As Ella Reeve Bloor's granddaughter recalls it, Ardenites survived thanks to "a lot of cooperation," growing food and canning it communally.[24]

In 1950, an additional seventy-four acres were added to the Ardens. Although Arden's constitution had long included an anti-discrimination clause, prohibiting preferences based on age, race, or sex, having this additional acreage prompted leaders to make efforts to be actively inclusive: They went to Wilmington to invite African Americans to move to Ardencroft. As new families did, Arden became the first town in Delaware to voluntarily integrate its school. After World War II, the nearby DuPont company expanded rapidly, and soon suburbs pressed against the edges of the unorthodox community. In that staid era, rumors and innuendo swirled that Arden was a bastion of free love, a nudist colony, an anarchist haven, and a communist stronghold—most of which had once been true but no longer was.

Contemporary residents are no longer beset by such rumors, nor are most single-taxers. The Ardens have incorporated, and now take state money for road maintenance. But aspects of Arden's early days endure. The land is still owned communally by the Arden Trust and leased to residents. And a strong sense of community continues to be forged through the many "gilds," now including a Dinner Gild, a Shakespeare Gild, a Gardeners Gild, a Scholars Gild, and a Poetry Gild. In fact, the Dinner Gild has ensured that communal meals continue. On Saturday nights from October through May, volunteers prepare a community supper, an event that usually draws eighty to one hundred thirty to eat together.[25]

Arden's longevity is unusual among utopian experiments, but I propose that its most lasting influence has come not through persisting, but through its impact on one of its early, part-time residents. In 1905, Scott Nearing, then a young economist teaching at the University of Pennsylvania, visited Arden—and was so impressed that he promptly rented the last available lot abutting the village green. For the next decade, he spent summers and weekends there. In his autobiography, Nearing devoted just a few paragraphs to his time in Arden, describing it as an "off-campus liberal education" and "an opportunity to combine country living, professional work, bread labor and rewarding associations." That brevity belies the deep influence his time at Arden had on him—and thus on subsequent generations of back-to-the-landers.[26]

While Scott Nearing didn't agree with Frank Stephens that the single-tax model could be scaled up to solve America's woes, the two agreed on most other things. They each had a commitment to pacifism, a willingness to break with their country over war, the belief that land is essential to freedom, and commitments to equality and to vegetarianism. During Scott Nearing's time at Arden, he refined many of his ideas about economics, social justice, and the good life. When he and his wife, Helen, became "simple livers" in the 1930s, they adopted many elements of the life Scott had enjoyed at Arden. The pair chronicled their efforts in *Living the Good Life: How to Live Sanely and Simply in a Troubled World*, which made the Nearings cultural icons in certain circles and inspired many of the hippie-era seekers who wanted to live more simply, freely, and authentically.

5

SCOTT AND HELEN NEARING
LIVE THE GOOD LIFE

A FEW DAYS PAST solstice in the summer of 2017, I packed my car and drove ninety miles to Harborside, Maine, to spend a week in the clapboard farmhouse where Scott and Helen Nearing lived from 1952 until 1977. During Scott's century-long life, he was part of the second wave of back-to-the-land experiments, during his time at Arden; part of the third wave, during the Great Depression, when he and Helen began homesteading; and a role model to many in the fourth wave, during the 1970s. The Nearings' *Living the Good Life* inspired so many in the 1970s that they've been dubbed the "grandparents" of that era's movement. And in 2014, the education director of the Maine Organic Farmers and Gardeners Association pointed to the couple's influence on the current generation of back-to-the-landers, saying, "You can draw a pretty straight line between the vibrance and the vigor of the sustainable food movement now and the seeds that the Nearings planted back then."[1]

On my way to Harborside, I stopped in Belfast to visit the United Farmers Market. The consortium of farmers, artisanal food makers, and crafters had opened just a few weeks earlier in a massive, metal-sided warehouse. I was hoping to catch up with

Kelby and Pam Young, from Olde Haven Farm, who were among the market's inaugural members.

A few years earlier, the Colorado couple were vacationing in Maine and became enchanted. As they drove up the coast on their way to New Brunswick, Pam began dreaming of homesteading somewhere along the Maine coast, envisioning a simpler, less hectic, and more purpose-driven life than the one they and their three children had outside Denver. By the time they'd returned home, Kelby was thinking the same thing; "I think we just looked at each other and were like 'We're gonna move,'" he'd told me when we first met. Unlike most people who have such dreams, they pursued theirs—even though neither had gardened, much less farmed. Pam and the children came east first, settling into a house they'd found online, building chicken coops, getting their first chickens and guinea hens and goats, and planting a vegetable garden. Kelby, an electrician, had to finish some jobs, so joined them a few months later. Within a year, the imagined homestead had morphed into a diversified farm.[2]

As I made a circuit through the market, I looked for the Olde Haven Farm logo and checked out what was available, an array that included lobsters, strawberries, hand-carved bowls, old photographs, granola, cheese, beet greens, and alpaca yarn. One vendor, Dig Deep Farm, had gorgeous lettuce; aware that the Nearings' old place is a good forty minutes from the closest grocer, I stocked up. Turning, I spotted Kelby two booths over. As we caught up, he bemoaned the cold, wet start to summer, which had killed his early crops. I sympathized; our peas were six inches high instead of two feet tall. He perked up a little as he described his pigs, so I bought two pork chops from him before getting back on the road.

Harborside is a tiny section of Cape Rosier, in the village of Brooksville, on the Blue Hill Peninsula in Downeast Maine. An hour and a half north of Belfast by car, it's actually a few miles southeast as the crow flies. On US Route 1, I passed a succession of fields and knolls studded with glacial moraines. The boulders and outcroppings looked as if they were nestling in the greenery—tall grasses, nettles, and milkweeds not yet in flower rose up to meet the rocks. Closer to Harborside were a few equally rocky blueberry barrens, some still tinged with last autumn's reds and rusts, others greening up and beginning to flower. The craggy, uneven fields don't look like a stereotypical agricultural landscape. Winding around the almost-island on rutted dirt roads, I passed below scudding clouds dumping rain erratically. No wonder so many New England farmers hightailed it west after the Civil War, happy to start over on flat, loamy plains where they could check the horizon and know what the weather would be.

Before they moved to Maine, Scott and Helen Nearing had homesteaded in Stratton, Vermont, for twenty years, from 1932 until 1952. When the area became overrun with skiers, they looked for a more remote spot to continue their Good Life experiment. Helen Nearing—who was mystically inclined—wrote that she dowsed over a map to find their new home and the rod repeatedly settled on the edge of the Penobscot Bay. Their second Forest Farm abuts this deep, island-studded inlet of the Gulf of Maine.

A small, easy-to-miss number on a telephone pole marks the farm's address. Doubling back after initially failing to spot it, I made out a lightly graveled driveway and followed it past meadows brimming with buttercups and daisies and thistle and vetch until it dwindled to grass. Amid tangled plants, I could see hints of garden walls. The Nearings always built enclosed vegetable gardens, with five-foot-high stone walls to keep out the critters. Most of this one was long gone, but the gate was still there, and beneath the weeds were one corner and some crumbling sections, enough to imagine its contour. On the driveway side, cadmium-yellow lilies and lemon-sherbet iris were fighting a losing battle against a phalanx of stinging nettle.

The farmhouse itself was tucked within a double cocoon of green. Around the edge of the cleared land, spruce, maples, locusts, and birch created a tacit boundary, ringing the half-mown, half-meadowed yard. Butterflies darted above a clutch of dinner-plate-sized peonies still clinging to their rightful place among the encroaching weeds. Closer to the farmhouse and outbuildings, a second arc of thick foliage—lilacs that had just gone by, gnarled apple trees, massive willows, and gangly bushes of beach rose—screened and shaded the house. Just beyond, an abandoned, fading-orange Chevy pickup was listing amid the lupine.

I'd arrived.

JOURNEY BACK TO THE LAND

MY TREK TO HARBORSIDE took a little more than three hours; Scott Nearing's took just shy of seventy years. Born in 1883 in Morris Run, Pennsylvania, to a financially comfortable family whose patriarch—Scott's grandfather—owned the local coal mine, Scott was a precocious and self-confident kid who readily absorbed the practical skills and knowledge his grandfather imparted. By his teens, he'd begun questioning why some people, like his own family, had plenty, while others didn't have enough. His sense that economic inequality was unfair was reinforced as he read works by Tolstoy and Henry George's *Progress and Poverty*.

Despite this early consciousness of inequality, Scott didn't consider himself radical; to the contrary, he said that "in a sane, rational, kind society, such as I thought I lived in, I would have been a loyal and devoted citizen. By temperament, I am not a rebel or congenital dissenter." Maybe he wasn't a rebel, but Nearing was interested in social alternatives from a very early age. When he was in high school, he wrote a paper about Fourierism in America, arguing that it failed not because of its lemonade-ocean notions, but because of outside forces.[3]

Scott completed a bachelor's degree in oratory at Temple University and another in economics at the University of Pennsylvania's Wharton School of Business. He stayed on at Wharton to earn a doctorate in economics, then taught there. In 1905, while still a graduate student, he began renting his lot in the single-tax enclave of Arden, in Delaware. There, he not only learned about the single-tax philosophy and the Arts and Crafts ethos of integrating art and nature into everyday life, he also encountered the writings of Karl Marx for the first time, gardened organically, and learned to build with stone. Nearing credits Arden as one of the three sources of "first-hand knowledge" that propelled him onto a radical path. He also became convinced one could enact radical beliefs by going back to the land—a conviction that became the keystone for what he called "the Good Life."[4]

In 1908, Nearing married Nellie Seeds, a feminist and educator, with whom he coauthored *Woman and Social Progress*. Calling "woman's capacity" the "great undirected force in modern society," the Nearings proposed harnessing it to reduce inequality. Within just a few years, however, Scott shifted his attention away from gender-based inequities and toward extreme economic inequality, seeing the latter as the biggest problem the United States needed to address. Whereas the single-tax advocate Henry George blamed private ownership of land for economic inequality, Scott Nearing blamed capitalism and consumerism. City families, unlike rural families, had to be mostly consumers, and wage earners struggled to cover their families' basic needs. This precarity practically forced low-income households to send their children out to work. Even with whatever extra income a child might bring in, poverty remained widespread, and many Americans struggled to have enough to eat.[5]

In 1914, in his book *Reducing the Cost of Living*, Nearing observed, "The contrast between the day laborers' life and the life of the well-to-do members of the community was never so great as it is today," for "the poor have grown richer—a very little richer, while the rich have grown richer at an unheard of rate." He was right. In 1912, John D. Rockefeller's fortune was about one billion dollars, and the median household's wealth was eight hundred dollars. Rockefeller was 1.25 million times richer than the average family. A decade earlier, the richest person had been 370,000 times wealthier than the average family: The wealth gap had grown more than threefold in just ten years.[6]

A personal solution to inequality, Nearing suggested, was moving back to the land. Doing so was, "in its essence, a protest against the intricacies and complex relationships of a society in which men have become so interdependent that they feel the yolk [sic] of enforced cooperation pressing upon them." He echoes Thoreau, who had lamented that the fellow-man to whom one is yoked is "a steer that is ever bolting right the other way." As urban demands further limited freedom, Nearing imagined "the city man" gazing longingly toward the countryside and seeing a farmer, who "is no man's man. He is free!" Nearing acknowledged that farmers also face constraints, but maintained that, in the "large sense, however, the city man is right. The farmer can, if he will, step out of the machine." By so doing, "men and women find in rural life freedom from servility."[7]

Urban inequality, the source of so much unfreedom, was due, in large part, to the increased corporate presence combined with a lack of minimum-wage laws. Only Massachusetts had one, then recently passed. Like other progressives in that era, Nearing urged passing minimum-wage laws, and basing them on three fundamental propositions: "1. Industry must pay a wage sufficient to maintain the efficiency of its workers. 2. Society must oppose any wage that leads to poverty, hardship or social dependence. 3. Wages must be sufficient to enable the worker and his family to live like self-respecting members of the community." He calculated that between 80 and 95 percent of wage workers in American cities earned less than they needed for a minimally decent life.[8]

He also advocated for laws protecting child laborers, which infuriated some members of the University of Pennsylvania's board of directors. They demanded that the university fire him, and despite being held in high regard by his colleagues and teaching the university's largest class, he was let go. A friend hired him to teach at the University of Toledo. Two years later, he was again fired. As had Frank Stephens, Nearing criticized President Wilson's decision to enter World War I, arguing it was motivated by corporate greed. Laying out that argument in a pamphlet entitled *The Great Madness*, he was charged with treason. Again like Stephens, he was acquitted. But he lost any remaining chance for an academic career. No universities would hire him, and his publisher dropped him—and his widely used textbook—from its roster. Soon, no commercial press would publish his work, either.

Scott Nearing didn't give up. For years after the dismissals, he argued for racial equality, economic reform, fairer labor practices, pacifism, and other progressive causes. By the 1920s, he recognized that those seemingly discrete issues were interconnected. In 1929, in *Black America*, he argued that the economic subordination of African Americans was not only an act of exploitation, but also a strategy on the part of white business owners to keep working-class Americans separated by race so they wouldn't unite in demanding better labor conditions. During the 1910s and 1920s, Nearing was a prolific writer, an active lecturer, and a political figure. He joined and then left the Socialist Party, joined and later was expelled from the Communist Party, traveled to the Soviet Union and China to see their socioeconomic systems for himself, participated in public debates against the likes of Clarence Darrow and Bertrand Russell, and even ran for Congress in New York (unsuccessfully) against Fiorello La Guardia and as the Workers' Party candidate for governor of New Jersey (also unsuccessfully). Believing, as so many other utopians had, that capitalism encouraged moral decline, Nearing was an ardent and seemingly indefatigable activist, publicly railing against its ills—until 1932, when he suddenly, seemingly, wasn't.

By then, both his personal and his professional life had been turned upside down. He and Nellie separated in 1925, and his relationship with his sons, then eleven and thirteen, became and remained strained. Nellie said she didn't believe in divorce, so the two were legally married until her death. But in 1928, Scott Nearing met Helen Knothe, a vivacious and unconventional woman twenty years his junior. The two quickly became captivated with each another.

Helen Knothe and her parents were theosophists, followers of a philosophy with some similarities to transcendentalism. Like the transcendentalists, the Knothes believed the universe itself to be a manifestation of the divine, making all its aspects divine and inherently interconnected. As it had for Bronson Alcott, this belief led the Knothe family to be vegetarians at a time when few Americans were. The two philosophies also had in common the belief that individuals can access the divine directly. (Theosophists hold some beliefs, such as reincarnation and karma, that aren't part of transcendentalism.) Perhaps because of these similarities, Helen became an ardent admirer of Henry David Thoreau.

During Helen's childhood, her family home was "a center for political and social interchange among Leftist intellectuals and musicians," prompting Helen to imagine "possibilities of a life of social significance, of involvement in social causes." Her main interest, however, was the violin; by her teens, she was an accomplished musician. After high school, she went to Holland to study violin with a renowned instructor.[9]

Two months after her arrival there, she met Jiddu Krishnamurti, whom many theosophists believed was the "World Teacher," a kind of messianic figure. Krishnamurti was a charismatic man and talented public speaker who attracted many attendees to Theosophical Society events. In November 1921, barely two months after they'd met, Krishnamurti declared his love to Helen. It took her two more years before she chose him over her violin, but once she did, the two had a significant relationship until early 1926, when they parted ways.

After that relationship ended and Helen moved back to the United States, it was to her family's home. She did some secretarial work for her father; in 1928, he asked her to contact Scott Nearing and invite him to speak at a social club gathering. Scott, in turn, invited Helen to join him on a long drive to do an errand, and Helen reported feeling almost instantly smitten with him. Helen and Scott seemed an unlikely pair. In addition to the age difference were temperamental ones—Scott was more rigid and austere; Helen was light-hearted. Her biographer, Margaret Killinger, proposed their many differences made the relationship work; theirs was "a synthesis of spiritual and political expertise, while also a union of a consummate teacher and inveterate student."[10]

As their relationship developed, Scott urged Helen to move into a tenement apartment and get a low-wage job to better understand how the other half lived. For the winter and spring of 1929–30, she rented an unheated cold-water flat and held jobs at a paper mill, a box factory, and a candy-packaging plant, moving from one to another when she and fellow workers were laid off. While immersed in this dreary situation, subsisting mainly on grapefruit and Triscuits, Helen learned Nearing's secretary was pregnant with his child. Scott insisted this wasn't true; however, in the wake of this disclosure, he urged Helen to return to Europe to figure out if she really wanted to forsake the high life for a "low life" with him.[11]

Helen left that summer. When she cheerfully wrote to Scott that she felt free as a bird, he sent a sharp retort, enumerating the reasons she was not, in fact, free: "In a world where the food and clothes you use daily are produced under the labor conditions that you sampled in NYC last spring," the letter began, "you cannot be free of an obligation to produce food and clothes," and then went on to enumerate her other unfreedoms. Apparently, neither his dalliance nor his imperious tone deterred Helen; when Scott sent a cable asking if she'd come back and work with him, she said she "could not resist him or the opportunity to do some serious work. I quit suitors and the high life . . . [and] off I sailed on the next available steamer, knowing full well that this was another turning point in my life." The two moved in together. "Poor as church mice," as she later recalled, they lived together in "the slums of New York," spending their days in the library doing research and writing and their evenings in a wood-heated three-room apartment at Avenue C and 14th Street, on the Lower East Side—one of the most densely populated sections of the city.[12]

Like many other urbanites, the Nearings spent almost half their income on food, more than they paid for rent, and struggled to find sources for fresh produce. Thanks to advances in food preservation technology, canned vegetables and meats dominated the market. The canned versions often bore only a remote taste resemblance to their fresh counterparts, but Americans in the early years of the twentieth century "crave[d] factory-processed foods." Among the many processed items developed in the 1920s still on the shelves today are Wonder Bread, Quaker Oats quick oats, Yoohoo Chocolate Drink, Reese's Peanut Butter Cups, Welch's Grape Jelly, Honey Maid Graham Crackers, Hostess cakes, Rice Krispies, Peter Pan Peanut Butter, and Velveeta cheese.[13]

Though the Nearings avoided processed foods, they couldn't escape the fact that people in poor neighborhoods often paid more for an item than did their wealthier counterparts. In upper- and upper-middle-class areas, customers could shop in one of the new "economy" grocery supermarkets, Piggly Wiggly or the A&P, the Walmarts of their day. But working-class communities were served mostly by food carts and small grocery stores that lacked the economies of scale chain stores enjoyed. Witnessing the lengths to which people went simply to feed themselves, Scott Nearing wrote a pamphlet entitled *Must We Starve?* He answered *Yes, we must*—unless the mainstream economy was reconfigured so profits would be split more fairly between laborers and owners.

But Scott was growing pessimistic about the likelihood of that happening. For liberals and radicals, the first two decades of the century had been full of promise: "Liberalism and radicalism in the United States were recognized as a normal part of public life. Up to this point, for decades, the Left had enjoyed a position among recognized American institutions. Differences of opinion were not only tolerated but extolled as a 'natural right.'" But that was changing; by the end of the 1920s, as he wrote, "controversial issues were termed subversive, against the public interest, unpatriotic, disloyal, even traitorous. Leftist speakers were neither encouraged nor permitted to air their views."[14]

In 1931, Nearing quit his roles in the Workers School and the American Civil Liberties Union, on whose national committee he served. He began to focus on individual transformation as the route to a more egalitarian society, much as Thoreau had. He imagined a simple life in the country would enable Helen and him to enact their ideals: They could be pacifists who no longer contributed to social or economic inequality or to a war-fueled culture. Plus, as he pointed out pragmatically, "if one is to be poor, it is better to be poor in the country than in the city because one can at least grow food instead of having to buy it from the barrows or pick it out of garbage cans on city streets." After looking for an intentional community to join, and not finding one that suited, Scott and Helen Nearing moved to a farm in Vermont.[15]

As she had trained to become a concert violinist, Helen had none of the skills needed to be a successful homesteader. However, she was a quick study, committed to their experiment and willing to adopt the regimented approaches Scott formulated. She soon became competent in the garden and mostly harmless in the kitchen. She also became an enthusiastic, skillful stonemason; the couple built more than thirty structures out of local stone, creating barns, outbuildings, garden walls, and a house in Vermont, then did it all again in Maine. Echoing Thoreau, Helen said she and Scott "dared to believe that ordinary, untrained people could make their own homes as birds build their own nests."[16]

STEPPING BACK

Unlike Thoreau, who had the skills to lead a simple, self-sufficient life, most Americans in 1932 would have been almost as ill-prepared for self-sufficiency as Helen Nearing was. For the first time in the nation's history, more Americans lived in cities than in rural areas. World War I had created more demand for factory workers than could be met by white laborers, which catalyzed the first significant wave of African American migration from southern agricultural lands to northern industrial cities, and most who'd come north stayed after the war ended. Overall, fewer than a third of Americans were still farming.

For some, city life held great appeal. Those in the middle and upper classes could enjoy new amenities like indoor plumbing, refrigerators, and electricity to power lights and radio. (Most rural homes, in contrast, had to wait until the end of the 1930s for electrification.) For financially secure urbanites, the outhouse and the chamber pot were things of the past, preserving food was easier than ever, and information and entertainment were available at the flip of a switch. That wasn't the case for the working class. Many still lived in densely populated tenements, sometimes upward of five thousand people in a single square block.

During the Great Depression, however, economic well-being eroded so much that destitute urbanites began turning to the land. As in earlier eras that had seen large waves of utopian experiments, this one included religious communities; among them were Protestant denominations, Jewish, Catholic, and Quaker groups, and the first Buddhist community in the United States. Black Mountain College was established, and both students and faculty were responsible for all aspects of living, not just of learning. Many other communes and collectives were aided by the New Deal, as part of the federal government's effort to keep people from starving. And, as the utopian scholar Timothy Miller noted, during the 1930s "the socialists, anarchists, cranks, visionaries, and other idealists . . . continued to make their contribution to the fabric of community. As ever before, intentional communities were centers of political agitation, social reformism, the creation of art, and cooperative homesteading." In addition, many back-to-the-landers set out on their own, as nuclear families of homesteaders.[17]

Among those "other idealists" was Ralph Borsodi. In 1920, a landlord sold the New York apartment Borsodi was renting out from under his family. The Borsodis moved out of New York and began homesteading. Unlike the Nearings, who avoided bank credit, Ralph Borsodi and his wife were proponents of any gadgets and machines that promised to make work more manageable, and willingly went into short-term debt to buy them. But, as Borsodi emphasized in *This Ugly Civilization*, they did so in service of their goal—to exit consumerism. Like so many of the utopian back-to-the-landers who preceded him, Borsodi believed factory production was "robbing the worker, his wife, and his children, of their contact with the soil." He wanted people "to be able to abandon the buying of the products of our non-essential and undesirable factories, and still be comfortable."[18]

To do that, he reasoned, "the home must be . . . an economically creative institution. It must cease being a mere consumption unit. It must become a production unit as well. It must be as nearly as possible an organic home—house, land, machines, materials and a group of individuals organized not for mere consumption but for creative and productive living." By 1936, Borsodi was confident enough in his own skills to open what he called a School of Living. He hoped artists, craftsmen, and teachers would come and learn to live self-sufficiently as homesteaders—sort of an academic version of Arden.[19]

Scott Nearing saw activism and self-sufficient living as arising from the same impulse. He wrote, "The advocates of simple living [and] the social radicals" initially do the same thing: "They criticize the mode of life in its entirety. [They ask] 'Why should one sex or color be subservient? Aren't we all human beings together in one world? . . . Must we always live in this horror? Why conform to such an inconsistent chaotic world?'" In response to these existential challenges, he said, both simple-livers and social radicals fight for change: "The simple liver devotes himself to one side of the problem, the social radical to another. The advocate of the simple life . . . fights for a free world, where the individual can express himself and join freely in life's experiences. The social radical combats injustice, exploitation and abuses." Surface differences aside, Nearing concluded that "the simple liver and the social radical are working toward that same end."[20]

In his early writings, Nearing mostly imagined the freedoms to be had on the land. In 1954, in *Living the Good Life*, he and Helen chronicled the freedoms, challenges, and joys they experienced while homesteading in Vermont. In the book's preface, the couple itemized their goals. They wanted to make sure people understood the perils facing the United States and wanted to be part of a resistance movement fighting against the "plutocratic military oligarchy that was sweeping into power in North America." They believed they had identified a way of "living sanely in a troubled world," and wanted to share it: "The ideal answer to this problem seemed to be an independent economy which would require only a small capital outlay, could operate with low overhead costs, would yield a modest living in exchange for half-time work, and therefore would leave half of the year for research, reading, writing and speaking."[21]

The "independent economy" they proposed—the "Good Life" of the title—was in no way easy or carefree. The Nearings were highly principled, committed to their vision, and almost entirely lacking in frivolity. When a reader of their "Mother Earth News" column asked what games they played, Helen answered with a tacit harrumph that she and Scott "don't play indoor or outdoor games. (Quite aside from being too old for them, we never did dissipate our energies in that way.)" While they thought the good life would involve working only half time, that turned out to be true only if they designated some necessary tasks—such as chopping the eight cords of wood they needed each year—as so pleasurable they didn't count as work.[22]

During my week at the Nearings' former farmhouse in Maine, I walked in their woods and thought often of Scott and those eight cords. I tried to envision seventy-year-old, then eighty-year-old, then ninety-year-old Scott cutting down a couple of dozen hardwood trees, with a friend or helper on the other end of a two-person saw, then cutting off the too-small branches, then lugging the trunks and big branches back to their yard. There, he'd have to chop them into log-lengths, split them, stack them to season—all without the aid of a chainsaw, gas-powered splitter, or skidder. Did it stay pleasurable? It must have. As he aged, he got some help with the heavier aspects of the task, but Scott himself split and stacked wood almost every day until he was ninety-nine years old.

As were earlier back-to-the-land utopians, the Nearings were animated by desires for freedom and equality. They shared their predecessors' belief that a good life required living with "other creatures who are also inhabiting this planet at the same time that we are, not exploiting them, not using them, not over-riding them, not preying on them, not eating them, not enslaving them, of course." And they regarded capitalism as a source of human suffering, making people unfree and unequal. By not eating meat, the couple freed themselves "from the slaughterhouse diet"; they also "resisted all attempts of the profiteers to sell [them] habit forming drugs," including "alcohol, tobacco, coffee, tea, or cola products."[23]

Likewise, they shared their forebears' preference for a use economy over a market economy, believing "a market economy seeks by ballyhoo to bamboozle consumers into buying things they neither need nor want, thus compelling them to sell their labor power as a means of paying for their purchases." Like the Arts and Crafters whose ideas suffused Arden, the Nearings believed commodity culture had "turned millions of talented humans into spectators who stand outside all the creative processes of nature and society and feel their own creative impulses shrivel and die." To counter that, their homesteading involved "living under conditions that would preserve and enlarge joy in workmanship." They took real pleasure in making, be it music or a stone wall.[24]

FOREST FARM, A NEW ARDEN?

ON MY SECOND DAY at Forest Farm, Warren Berkowitz stopped by to say hi. Warren runs the Good Life Center, a nonprofit based in the stone house the Nearings built in the 1970s—their home for the final years of their lives. As do Fruitlands and Walden and some of the other sites where utopians have lived, the Good Life Center perpetuates the ideals and legacy of its former inhabitants. Each summer, a gardener-in-residence cultivates vegetables in the Nearings' beautiful stone-walled garden and welcomes the two thousand visitors who come to learn about the Nearings. On Monday nights during the summer, the center continues the tradition the Nearings established of hosting gatherings in the large living room to discuss important issues.

I'd met Warren during my first visit to the Good Life Center and immediately liked him. He is soft-spoken, attentive, and warm; a wispy ponytail and wire-rimmed glasses evoke his hip youth. During that initial visit, after letting me spend several hours in the Nearings' library, he invited me to his home to meet his wife, Nancy, who was equally kind. When he stopped by Forest Farm, we sat at a plain, quarter-sawn kitchen table. He remembered it from the year, back in the late seventies, when Nancy had lived in this house. His eyes roamed the room.

"Maybe something's different," he said, "but it looks exactly like I remember it."

The cupboards are graced by hand-painted Dutch figures that were there before the Nearings bought the house. I bet they'd delighted Helen, who was of German-Dutch ancestry and had loved her time in Amsterdam. The woodstove, dish rack, and even the steel sink combo were all there in the seventies. I knew of only one thing that was definitely different: A DSL router sat blinking inside one of those jauntily painted cupboards, plugged into the only outlet in the house that accepts grounded plugs.

Because I'm hopelessly digitally tethered, I'd confirmed that the house had Wi-Fi and cell service before signing the rental agreement. I hadn't thought to ask if it would be *good* service— and it wasn't. To use my cell phone, I had to sit by the sink, find the service sweet spot, and stay almost perfectly still. Internet service was far worse; I could get to a browser but couldn't load pages. Within a day, I'd abandoned Wi-Fi in favor of a cell-phone hotspot, promptly burning through my entire data plan. Aware of the Nearings' aversion to "clutter and gadgets," their umbrella term for telephones and radios and televisions, I chalked up the lame service to otherworldly interference by the psychically in-clined Helen.

Not that I was at a loss for things to do. Every day I had people to see, and plenty to read in the evenings. I'd brought *Living the Good Life* and Helen's cookbook, *Simple Food for the Good Life*. I also had digital photographs of a manuscript for "Arden Town," a novella Scott wrote in 1913. I'd made the photos with my cell-phone camera a few weeks earlier, when I'd gone to the Walden Woods Project to look through Helen's and Scott's archive. Al-ready I had a sense that "Arden Town" was more a polemic than an absorbing fiction; still, I was excited to read it. Over two evenings, with only a forty-watt-bulb lamp to augment the waning light, I read the story on my cell-phone screen. What struck me in that first read was how every important character was a version of Scott Nearing—it was like the scene in *Being John Malkovich* when Malkovich goes through the portal in his own head and every person he meets is a projection of himself.

The protagonist of "Arden Town" is Fred Allen, "a brilliant young business man—full of noble impulses." He's got Scott's looks. Arden's philosophy (which is, in large part, also Scott's) is channeled by the suffragette Ruth Wade, "an artist—a girl of twenty five." Another Ardenite, Sam Martin, has Nearing's vocation and avocation: He's an economics professor by training and an enthusiastic organic gardener. Will Evans, "the spirit of the place," is based on Frank Stephens; he propounds the ideals about ethics, work, and craftsmanship Nearing also held.[25]

When "Arden Town" opens, businessman Fred Allen is instructing his factory foreman to incorporate "kiddie labor" to cut costs. Fred and an old friend are about to leave the city for a week at the shore, where Fred expects to make official his all-but-engagement to the socialite Miss Treat. But his friend's cousin Ruth joins them on the train ride, and Fred finds her fascinating. She schools Fred in macroeconomics, the need for women's suffrage, and the importance of maintaining a commitment to aesthetics in all aspects of one's life—even in business. Afterward, the pragmatic businessman says, "She has turned me inside out . . . but I am not at all sorry."[26]

A few days later, Fred and Ruth head out for a sunrise sail, and Fred asks her to tell him more about Arden. "It is suffused with a wonderful atmosphere which makes the grass a little greener, the sky a little bluer, and the stars a little brighter than they are anywhere else," she tells him. "One feels Arden." Ruth adds that there "we grow in all ways into nobler selves." Fred expresses doubt, asking, "Aren't people just the same there as they are anywhere else?" Ruth assures him they are, that people are "naturally decent and honest. They are always that way if you give them a chance to be."[27]

She attributes the neighborliness of the residents to Arden's adherence to Henry George's single-tax philosophy, which doesn't make sense to Fred. "No one can understand until he has been there," she tells him. "Everyone expects everyone else to be honest, and they are—that is all. It is the spirit of the place . . . Each tries to do as much as he can for the rest, and the whole place is a beehive of industry—cooperative, mutually helpful. We are all on an equality in Arden." Lest the reader miss this point, Nearing has Fred ask whether Arden is heaven. "It is near heaven," Ruth says. After they come ashore, Fred hammers it home: "What you say is Utopian, but it is also inspiring and cheering."[28]

Fred asks if he can come to Arden and three weeks later, he does. There, he falls in love with Ruth, of course. More important, he falls in love with Arden—he discovers that Ruth hasn't exaggerated the sense of mutual helpfulness. Just about every activity is communal: The schoolteachers, the parents who run the day-care center, the musicians for the dance, the participants in the upcoming pageant, the costume makers and set designers, all are Ardenites. As Ruth explains: "No one is ever hired here. We do it ourselves."[29]

In the novella's final sections, Fred is willingly corralled into helping with assorted tasks. He makes posters for the pageant, assists in moving a piano, repairs a doll, weeds strawberries, helps lay a floor, moves a stone pile, strings up Japanese lanterns, and gathers firewood for the weekly campfire. As Fred is still in town on Thursday night, he has a chance to hear Sam, whose nickname is "the professor," give his weekly economics lecture—something Scott Nearing did at Arden in real life.

Fred is awed by Arden, sees it as unique among "experiment[s] in human brotherhood. Living close to nature, and close to one another's hearts, the Ardenites stood in a relation of peculiar beatitude toward mankind and toward life." Still, when Monday morning rolled around, Fred went back to the city, and didn't return to Arden until Halloween. During that later visit, he finally talks to Will Evans about moving to Arden, telling him "you have something here for which I have [longed] . . . The brotherhood, the spirit of service, the joy of human work and human play—you never find those things in the city."[30]

"Arden Town" sounds utterly implausible. But when Scott Nearing lived at the real Arden, it was quite like the fictional version, right down to the work parties, vegetable gardens, democratic government, and assorted entertainments. On Saturdays, residents could count on a Shakespearean reading or full performance in the amphitheater. On Sundays, they held a communal campfire where someone might read "a poem or selection, another may give a flute, violin or clarinet solo. There is chorus singing by the whole assembly." Baseball games, lectures, skits, readings, and elaborate pageants rounded out the recreational roster. One pamphleteer of the era wrote, "Arden is the capital of the State of Uncritical Friendliness; it belongs to the Federation of Mutual Helpfulness, under a Constitution of Equal Opportunity."[31]

A 1911 piece about Arden for *Suburban Life* magazine singled out Nearing. It described him as "a clever reader of the classics, and at the Wednesday evening lectures he takes an active part, besides being an all-round athlete and worker in numerous ways. He leases several acres of ground, and has proven, without doubt, that agriculture is no lost art, being a good farmer, as well as a keen and active student of this and many other industries." Indeed, during his time at Arden, Nearing was considered "one of the leading lights of the place, following close in the footsteps of Frank Stephens."[32]

When Scott and Helen Nearing moved to Vermont, they'd hoped to create something akin to the sociable, communal Arden by creating a social and economic collective among the residents of their mountain valley. But in the closing pages of *Living the Good Life*, they concede it didn't happen. "Considered in terms of individual health and happiness," they wrote, "our project was an emphatic success. Viewed socially, however, even on its economic side, it left much to be desired." Their Yankee neighbors were individualists with whom they had less in common than they'd anticipated. And few formal mechanisms for fostering new community existed; the handful of gathering spots frequented by locals promoted behaviors the Nearings eschewed, like drinking alcohol and eating doughnuts.[33]

Though uncommonly capable, the Nearings could have used more help. If only their "group consisted of a dozen or a score of capable adults, animated by the same purposes and willing to follow out agreed plans, along lines of well-established practice," they wrote, "our standard of living could have been attained and maintained with far less expenditure of energy and labor time, leaving much more time and energy for leisure and avocational interests." Even with additional comrades, they doubted they'd have had "enough good voices to make up a choir, nor enough passable musicians to man a local orchestra or enough actors to set up an effective dramatic troupe or dance team," the pastimes enjoyed in both the real Arden and the fictional Arden Town.[34]

I hoped the couple had fared better in Harborside. They wouldn't have found many kindred spirits among the long-term residents; wary locals dubbed Cape Rosier "Commie Town" after the Nearings moved there in 1952. The postmistress reported their comings and goings to the FBI. But so many young idealists and counterculturalists came to meet them, especially during the 1970s, that it seemed possible they'd have found enough voices for a choir. I would have the chance to find out soon, as Warren and Nancy Berkowitz had invited me to come over on Wednesday and I was seeing Eliot Coleman on Thursday. All three had moved to the area because of the Nearings. But first, I needed to prepare for lunch guests who were coming on Monday.

COOKING IN HELEN'S KITCHEN

THOUGH SCOTT AND HELEN Nearing believed in gender equality, they still defaulted to a variety of stereotypical roles, as Helen cleaned the house, cooked the meals, managed the household budget, and took care of their clothing. Scott's dietary preferences and her lack of interest in cooking led Helen to make very basic meals that could be eaten out of wooden bowls, their only tableware. They frequently had "mono-meals," just a single food, and they usually fasted once a week, including on Thanksgiving, when most Americans were feasting (or, as Helen called it, "glutting").

Despite Helen's lack of interest in food preparation, friends and even her publisher urged her to write a cookbook. She proposed an "anti-cookbook" instead, to which, as she tells it, her publisher consented, "as long as it has 'cookbook' in the title." The first time I met Nancy Berkowitz she'd told me Helen considered mashing potatoes too much work, so I was hard-pressed to imagine what her cookbook would contain. Curious, I bought a copy anyway. Along with quotes from other cookbooks and diatribes about processed food, the book consists of simple recipes requiring just a few ingredients. Helen dawdled a good long while before getting to the recipes, though; the first one is on page 88.[35]

Two friends were coming to visit me at the Nearings' farmhouse, and I wanted to make our lunch from Helen's cookbook. Wheat berries, rolled oats, and popcorn were the Nearings' non-vegetable staples. I didn't have rolled oats and didn't want to serve popcorn to guests, so wheat berries it would have to be. A day ahead, I set out two cups to soak overnight. In her cookbook, Helen recommends drinking the soaking water the next morning, after draining the berries. I took a sip; the flavor was vaguely nutty and a little tangy, but off-putting. I poured the rest down the drain. After preparing the wheat berries two ways, one with honey (Scott's favorite), the other with olive oil and sea salt (Helen's favorite), I made a salad, a dish well represented in the cookbook. Helen liked adding a bit of fresh basil, which was in season, so that's what I did. Water—fresh from the tap, not some that had been sitting out overnight—would be our main beverage.

I got a kick out of making Helen's recipes in her kitchen, but the conscientious host in me needed a backup plan in case the wheat berries were gross, so I added several things Helen would have shunned: a cheese board, rosé, and whole-wheat rolls. As much as the Nearings spurned alcohol, I suspected the rolls would've been the final straw for Helen. She was so anti-bread that her chapter titled "Baked Goods & Over-Starching" opens with a dismissive quote from a sixteenth-century cookbook: "Much dough-bake I praise not . . . if I had two loaves of bread I would sell one to buy a hyacinth to feed my soul," to which Helen added, "If I had two loaves of bread I would sell them both to buy two hyacinths."[36]

Helen concedes she'd had a few tasty loaves, including "the black *chorni hleb* of the Russians, the *roggebrot* of the Dutch and the coarse sourdough bread that our grandson-in-law, Peter Schumann, director of the Bread and Puppet Theater, makes and distributes free at his performances. All of these are made of un-bolted and unrefined rye." There it was again—unbolted flour. Helen railed that "the modern method of processing—purifying (deleting the bran), fine grinding, fine sifting, bleaching, 'enriching,' adulterating—robs good cereals of their life-giving energy. Bread becomes not the staff of life but the crutch of constipation, illness and death." The section could have been taken straight from Sylvester Graham's harangues from a century and a half earlier. Why, all those years later, were people still ruining staple grains? Helen's answer: "a dead flour is required for bulk storage." She's right; denaturing it makes it shelf-stable.[37]

As both Helen and Scott knew, many foods were diminished by excessive processing and by adulteration. Scott probably learned way more than he wanted to about food dangers from his Arden neighbor Upton Sinclair. In fact, he specifically cited "the menace to health arising out of food processing and poisoning and a determination to safeguard ourselves against it" as one of the "chief factors that took us out of the city into the country." He and Helen decried the food processors and chemical-additive companies, saying they "deliberately devitalize, drug, and poison the population for profit. Perhaps it may seem absurd, in this day and age, to write about deliberate poisoning. Most people associate poisoning of food with family feuds in the Middle Ages ... [but] research shows the words are more applicable today."

These poisoners "produce about three thousand products which are used to preserve, flavor, and color food, and otherwise make it more attractive to the customer and more profitable to the merchant and his business associates. Among these chemicals are some of the most deadly poisons ever produced by human ingenuity." Despite the accomplishments of Dr. Wiley, the Poison Squad, and Upton Sinclair, Nearing's assessment had much truth to it.[38]

In 1933, the year after the Nearings moved to Vermont, two consumer-rights advocates, Arthur Kallet and F. J. Schlink, published *100,000,000 Guinea Pigs: Dangers in Everyday Foods, Drugs, and Cosmetics*, a grim account of the risks those chemicals posed. The authors argued that corporations routinely released products they knew might be harmful, and the Food and Drug Administration didn't always adequately police them because it was beholden to special-interest groups. Surely some members of the FDA looked the other way, but other FDA officials were quite principled. Two of them created a traveling show, also in 1933, that featured items the FDA knew were dangerous or useless but were powerless to ban. Nicknamed the "American Chamber of Horrors," the show highlighted medicines, foods, and cosmetics.

Some were merely fraudulent, like Bred-Spred. Advertised as a jelly, Bred-Spred didn't contain the 50 percent or more of fruit or juice legally required to market it as such. It contained no fruit whatsoever, and was made of artificial colors, artificial flavors, pectin and "a few hayseeds thrown in to simulate strawberry seeds." Other products were dangerous: Dinitrophenol, for example, was advertised as an aid in weight loss; it caused deadly blood disorders and cataracts but couldn't be regulated by the FDA because it was marketed as a cosmetic, not a drug. Othine promised to remove brown skin spots, but with mercury as its main ingredient, it also caused neurological and pulmonary damage, renal damage, cognitive impairment, and, sometimes, death. The "American Chamber of Horrors" did what its creators hoped: Shocked viewers began advocating for stronger, more sweeping regulation. But once again, Congress was slow to act.[39]

Much as Sinclair's The Jungle provided the nudge that got the 1906 food-safety laws passed, the "Elixir Tragedy of 1937" pushed Congress to pass the Food, Drug, and Cosmetic Act. The pharmaceutical company Massengill released a sulfa drug dissolved in diethylene glycol, which almost immediately killed 105 people. Diethylene glycol should never have been used in that way; it's a powerful industrial solvent and an ingredient in brake fluid. But the FDA didn't find Massengill at fault—because the ingredients in the product matched the ones the company said were in it. As long as they were correctly listed, deadly ingredients were legally permissible. The new legislation closed that absurd loophole.

Naturally, Scott and Helen Nearing were horrified by such deliberate, greedy behavior. They were also disheartened that even wholesome foods were less so when they reached the city. "City markets cannot sell fresh food," Scott wrote, as "the time span between producer and urban customer must be at least hours and days; at most it may be weeks, months, years." Like Thoreau averring "a huckleberry never reaches Boston," for its "ambrosial and essential part . . . is lost with the bloom which is rubbed off in the market cart," and like contemporary local food advocates touting seasonal foods grown nearby, the Nearings knew foods transported long distances and stored a long time weren't as good as fresh, local options.[40]

With abstinence as the cornerstone of their dietary decisions, the vegetarian Nearings bypassed meats prepared under the nightmarish conditions Sinclair documented. Avoiding processed foods enabled them to be sure they weren't ingesting potentially poisonous additives and adulterants. Growing food and eating it as close to raw as practicable, they minimized exposures to food-borne risks. Simple as the Nearings' diet was, and as indifferent a cook as Helen was, everyone I've met who ate at Forest Farm agreed that what the meals lacked gastronomically was more than made up for by the company. My Forest Farm lunch with friends followed that pattern. The wheat berries with oil and salt were sort of okay (the honeyed version less so). But conversation flowed and no one left hungry—a successful meal.

FOREST FARM HOSPITALITY

WHEN THE NEARINGS LIVED at Forest Farm, they had so many visitors they eventually posted a sign at the end of the driveway: VISITORS 3–5, PLEASE HELP US LIVE THE GOOD LIFE. Not everyone honored that request, however. Many misestimated how long it would take to get to the remote location, and others just showed up. A dozen visitors a day was common; once, Helen said, three hundred people came on a single day.[41]

Visitors could help Scott Nearing with whatever he was doing, but he rarely stopped to chat. It fell to Helen to give folks the nickel tour—showing them the house, the garden, their library, the composting system. Like Thoreau learning to play host at Walden Pond and the Fruitlanders and Brook Farmers performing for curiosity-seekers, Helen cultivated a public persona for the strangers who stopped by. Nancy Berkowitz recalled that people "would write and say 'I want to come on Tuesday at one o'clock.' She would say it was okay for them to come at one. The next set of people would want to come on Tuesday at two fifteen, some at four thirty." They weren't people Scott or Helen knew; rather, "they read their book, they saw them at a meeting, they heard them speak, they read their article in *Mother Earth News* and just asked 'Can we come?'" Not that Helen minded, Nancy added. To the contrary: "Helen loved to race people around. She'd say 'Come out to the garden.' You'd run through the garden; you'd run through the house; if she liked you, she would sometimes take you up on the balcony and show you the books; then out the door—in time for the next group."[42]

At Walden Woods, in Helen's archive, I'd found lots of old snapshots of the Nearings and their visitors eating outside. Usually, they were gathered in the dooryard between the barn and the farmhouse, with Scott sitting at a small table laden with serving bowls and Helen standing beside him. The guests sat on benches, low stone walls, or the ground. Nancy wasn't surprised by how many such pictures there were; she said Helen often invited visitors to lunch, "or sometimes they'd be having lunch, and somebody would just come, because out there—you saw how far it is, and how hard it is to get there—there's no other things to do on your way out. She'd get out a bowl and you'd eat with them. She did that up until the day she died."

Helen understood that part of her role was to *perform* utopia, to welcome the curious and show them what the Good Life looked (and felt and tasted) like. So perform she did; as one neighbor put it, "[S]he was almost like an actor in a way, almost like a larger-than-life performer amidst her own play. You know, 'This is what I'm doing, this is the good life, this is how you do it.'" Helen said she was trying to identify "possible recruits for a general effort now underway to stabilize and improve man's earthly living space." Whether or not they were recruited to the general effort, the young people who came usually helped out. Those strong young backs were a boon; though incredibly hale for their ages, Scott was eighty-seven and Helen was sixty-six when *Living the Good Life* was rereleased, by Schocken Books, in 1970.[43]

But like those who found fault with Thoreau because his mom did his laundry, plenty of admirers were disappointed that the Nearings didn't live up to their mythic image. They had financial resources they rarely spoke of, received construction help from friends and visitors, and hired professionals whom they rarely credited. Some felt betrayed by these elisions, especially young aspirants who wondered why they had such a difficult time homesteading when the much older Nearings made it seem manageable. Knowing the Nearings didn't rely on maple sugar or blueberries for their entire income, or that they didn't do everything themselves, would have been very helpful to new homesteaders. But it's also the case, as their neighbor observed, that they were expected to be larger than life. Some admirers put the couple on a pedestal so high that even small deviations from their stated values, like Helen's love of ice cream and her pleasure in having cats around, were cast as failings.

Although some lost their faith in the couple, most visitors remained admirers. And with a handful among the many thousands who came to Harborside, the feeling was mutual. The Nearings described most of the young enthusiasts who visited as "wanderers" who were "unattached, uncommitted, insecure, uncertain human beings." Among them, though, were "those serious few who are consciously and conscientiously working toward an ideal in which they believe and to which they attach themselves." Nancy Berkowitz was surely the latter. She met the Nearings in 1975 at the World Vegetarian Conference, came to Harborside for the summer in 1976, began working with Scott in 1977, and became practically family.[44]

"Many young couples, singles, and groups came to us wanting land," the Nearings recalled. Sharing Henry George's belief that land should not be used to create profit, the couple said they "let young people have the land at what it cost us." Though that's not 100 percent accurate, they did make it affordable. Most of it was sold in four lots to hippie-era back-to-the-landers. The first went to Eliot and Sue Coleman, who had discovered *Living the Good Life* at Hatch's Natural Foods, a Vermont health-food store with a lending library, and become captivated by the Nearings' lifestyle. In 1968, the Nearings sold sixty acres to the Colemans.[45]

Although Eliot Coleman said the "property wasn't necessarily what I was looking for [because] the soil was horrible, and it was covered with rocks and trees," the adventurous couple were glad for the chance to live and work near their role models. Dubbing their place Greenwood Farm, Coleman built a sturdy four-hundred -square-foot house using only hand tools—much as Thoreau built his cabin and the Ardenites constructed their houses. Appreciating the ethos behind the Nearings' decision to sell him land for just thirty-three dollars per acre, Coleman has since done the same thing, selling pieces to a few friends because he thought he "shouldn't be the only one receiving a gift." Leaving Maine for a decade after he divorced, Eliot Coleman returned to Harborside in 1990, and Four Season Farm has been going strong on that original parcel since.[46]

The Nearings offered Jean and Keith Heavrin, another back-to-the-land couple, acreage adjacent to the Colemans' plot. In May 1972, the Heavrins began transforming their forested parcel into a homestead. Jean remained amazed at their good fortune: "[Y]ears later," she wrote, "I asked Helen why she and Scott had picked Keith and me, out of the thousands who had come through their lives in those times, to share their land. She said she couldn't explain it to me, it was just something she and Scott had seen in us." The Nearings also sold a small plot to Greg Summers and sold their old farmhouse, the one in which I stayed, to Stan Joseph.[47]

Stan Joseph grew food, planted willows to use in making cor-
acles, and added many, many flowers to make into wreaths, including the lush layer of plants that still surround the house and
the irises and lilies near the now crumbling garden wall. Beside
the pond Scott had created by digging out and carting away fourteen thousand wheelbarrow loads of muck, Stan Joseph built a
sauna. He was infamous for photographing his friends, post-sau-
na, in all manner of undress. Inside the house are other remnants
from his time there, among them photographs, though none are
of the subgenre my photography students used to call "me-and-
my-naked-friends pics."

POPULATING PARADISE

LIKE JEAN HEAVRIN, I wondered what led the Nearings to select
this handful of people to receive what so many wanted. Helen
described the land as going "to the right young people," folks who
seemed "promising." She'd read their palms, seen something pro-
pitious. But surely others also had excellent lifelines or love lines
or contented-quasi-ascetic-who-will-work-from-dawn-to-dusk-and-
sometimes-after lines. She did say Eliot Coleman had the deepest
fate line she'd ever seen, so there's that.[48]

One thing Helen expected was that they'd all have children.
Sue Coleman was pregnant when she and her husband established
their homestead, and Helen was very direct in telling Jean and Keith
Heavrin she hoped they'd have children who could play with the
Colemans' daughter. When Stan was between girlfriends, Helen
made a point of asking friends if they knew any suitable women
she could introduce to him. Perhaps her preoccupation was due to
disappointment about not having children of her own; or perhaps
it was a way to secure more "recruits to the general effort."

Halfway through my back-to-the-land week, I visited Nancy
and Warren Berkowitz. They live about twenty-five miles from

Forest Farm, in Blue Hill, and I arrived in time to meet their daughter and son-in-law. The young couple were expecting a baby in just a few weeks, and as we all chatted, Nancy and Warren spoke tenderly about the prospect of becoming grandparents. The first time I'd met Nancy, she told me she'd worked with the Nearings for nineteen years. I was curious to hear her stories, eager for the perspective such a long intimacy could provide. But I also knew what might seem like a fascinating anecdote to me was her lived experience of an important relationship. That awareness was heightened considerably when she said she turns down most interview requests, explaining simply, "It was my life, my personal life."

In that moment, I felt especially forcefully the role of the living, loving, personal, and intergenerational connections I'd been tracing. As much as they were inspired by a yearning to be free, or a craving to be in nature, or a desire to reject consumerism, or a need to grow nutritious food, these movements were fueled by love. Whatever its shortcomings, the universal love promoted at Fruitlands was meant to animate people's lives, as was the quirkier model Fourier outlined. Love prompted Thoreau to go to Walden to write a book in memory of his brother, John, who'd died a few years earlier. Love daisy-chained into being a utopian community unbound by space or time, made up of people from different movements and generations who loved one another. Thomas Carlyle and Ralph Waldo Emerson, Bronson Alcott and Henry David Thoreau, Thoreau and Helen Nearing, Frank Stephens and Scott Nearing, the Nearings and Nancy Berkowitz are all connected as much by love as by ideas.

This recognition, I think, was at the heart of my obsession with figuring out if the Nearings managed to create their own version of Arden. The observation that the Vermont experiment "left much to be desired" downplays the depth of disappointment the author of "Arden Town" must have felt. For all its didacticism, the novella is awash in love—for the community, for the collaborative spirit, for the intellectual richness, for the place it-self. When Scott was fired from the Wharton School, he lost not only his job but also his home away from home at Arden. I'd been looking for evidence that he and Helen had made or found a community he could prize as much as he'd once cherished Arden. I was hoping Nancy would tell me he had.

After I prattled on at some length about all this, Nancy re-minded me that first and foremost, Scott Nearing was a teach-er. He and Helen lectured all over the world. They opened their home to a couple of thousand strangers a year who wanted to see them living their Good Life. They held weekly salons. As more and more people made their way to Harborside, folks built a campground between the Colemans' place and Forest Farm where visitors could stay, a makeshift community in the woods. A smile played across Nancy's face as she remembered the scene: "Lots of us liked living in the woods without anything. Helen and Scott were the hub. They had the electricity. We would come down and listen to music on Sunday nights. On Monday nights, we'd have political talks and neighborhood talks."

Nancy's emphasis on Nearing as a teacher brought to mind two things Helen told a reporter. The "cruel thing" about Scott be-ing drummed out of the academy, she'd said, "was that more than anything he wanted to teach one class with the same students and watch them grow." Then, just a few minutes later, she expressed dismay that they hadn't remained close to the Colemans or to the Heavrins or to Stanley Joseph: "We never see them."[49]

I think Scott hoped those young homesteaders would be his "class with the same students," the ones he could teach and watch grow. Hoped the community and camaraderie he cher-ished at Arden would imbue the scene at Harborside, not in the form of pageants and baseball games, but through the experiences he, Helen, and the homesteaders shared. But Stan Joseph was temperamentally very different from the Nearings, and both the Colemans and the Heavrins divorced, prompting first Eliot and later Jean to leave Harborside. Although he's been immensely influential, inspiring thousands of seekers, Scott was only able to keep his "class with the same students" for a short time.

Helen proposed that the rift between the Nearings and their neighbors was because she and Scott were too organized and methodical for the younger folks. I don't know, though it's true they were uncommonly organized. On my first visit to the Good Life Center, that season's resident gardener bypassed the garden and the house and led me straight to the compost, where he regaled me with details about Nearing's composting method. Apparently, Scott usually had between twelve and twenty compost piles. He kept meticulous notes about when each was started, what went into it, when he started using it, whether it was acidic or basic, always striving to add more fertility to the soil than had been taken from it.

If I did this, I could tweak the nitrogen or calcium levels available for different vegetables so they each had the optimal amount. But I know I won't; I'm happy simply making usable compost. And even though I've met many a great farmer, only a few strike me as committed to achieving quite that level of precision. One of them is Eliot Coleman, who is so conscientious about what goes into his compost that he describes it as "another mouth to feed." In many ways, he's Scott Nearing's successor.

FOUR SEASON LUNCH

ON THURSDAYS DURING THE growing season, Eliot Coleman and his wife, Barbara Damrosch, hold a leisurely crew lunch for everyone working at Four Season Farm. When I told them I was staying at the old Forest Farm, they kindly invited me to take part. Unlike Helen, Barbara is a great cook who enjoys making food; lunch was so delicious that as soon as I got home I bought the cookbook she and Eliot had cowritten. Their recipes are vegetable rich, but not vegetarian. After being vegetarian for a while out of respect for the Nearings, Eliot again eats meat. Though that might have offended Helen, who called omnivores "carcass eaters," she'd likely be pleased to see that their cookbook doesn't have any bread recipes. Which isn't to say they don't eat bread; two big loaves from Tinder Hearth Bakery were part of the lunch buffet.

At crew lunch, the group gathered around the table consisted of Eliot and Barbara; Eliot's daughter, Clara, who manages the farm now; four farmhands ranging in age from late teens to late forties; and me. Eliot, then closing in on eighty, was paying forward back-to-the-land wisdom, teaching folks young enough to be his grand- and great-grandchildren how to live simply, sustainably, as stewards of the land. He impressed on them the importance of soil, first and foremost, of creating a system so healthy that it can't help but produce nutrient-rich food.

Conversation zigzagged between Eliot's early adventures on this land and life now. He recalled the incredible sense of empowerment he felt the first few springs knowing he had enough food in the root cellar to survive until things started growing again. The idea of freedom came up frequently: Eliot insisted a sustainable farming/homesteading life is one of the freest imaginable, as it leaves you beholden to no one; Clara pointed out that the farmer may not be beholden to other people, or to the government, but is utterly beholden to the land, the crops, the livestock. The rest of the crew and I swiveled back and forth, from father to daughter, as they volleyed competing notions of freedom, of freedoms from and freedoms to, over spoonsful of ice cream.

All over the Blue Hill Peninsula, other farmers, artisanal food makers, and simple livers could well have been holding similar conversations. Having found their way to this remote spot because of the Nearings, or because of Coleman, or by sheer good luck, those seventies-era arrivals have been joined by members of the newest generation of farmers and homesteaders determined to eke out a beautiful life in this rugged place. Blue Hill is home to an uncommonly vibrant, forward-thinking farm and food scene. Brooksville, one of the villages, has only 934 residents but more than twenty farms and other food-centric enterprises "raising organic produce, livestock, oysters, and more, and making everything from preserves and goat cheese to bread and wine."[50]

Among these new producers are Tim Semler and Lydia Moffet, both children of seventies-era back-to-the-landers. In 2007, the couple started Tinder Hearth Bakery with the hope of creating not just bread and pizza, but also a communal space and community resource. In its early years, the bakery seemed like a throwback to an earlier age, with a gaggle of young adults living collectively, making awesome bread, and moving on after a season or two. Fun though that was, a larger ideal inspired the couple. Semler said their dream was to create community in the remote, rural area, adding. "It takes a long time." Moffet hoped Tinder Hearth would nudge folks in "some direction that could be helpful for future generations in some way."[51]

As the bakery gained traction, Semler and Moffet—with the help of family and friends—raised enough money for a real brick oven, which they built by hand. Still a groovy business with a great vibe, the bakery component of Tinder Hearth Bakery has exceeded the couple's early hopes. In 2017, *Bon Appétit* named it one of the top three pizza places in America; nowadays, customers call in their pizza order in the morning to be guaranteed a pie that night.

The current scene hasn't arisen by chance. Rather, there's "a definite strategy to it," according to John Altman, who began reviving the old farm David's Folly with his wife. Like Semler and Moffet, Altman knows making a resilient alternative to the mainstream requires knowing how "to cultivate the community" as well as the land. These efforts are succeeding; this remote peninsula supports three weekly farmers' markets during the summer and one in winter. And Blue Hill is at the forefront of a food sovereignty movement: Proponents maintain that farmers should have the right to feed their neighbors with minimal government regulation. In 2017, the Maine State Legislature approved a law permitting municipalities to craft their own food regulations, the first statewide food sovereignty law in the nation, and in 2021, voters approved a referendum to add a "right to food" to the state's constitution.[52]

During the 1990s, when Helen lamented the general lack of interest in going back to the land, Eliot Coleman used to remind her that it took forty years before his generation took up the Nearings' mantle. *It'll happen again*, he'd promise, predicting it would come again when the economy turned downward. "It was lost in the '20s—money, money, money," he said. "It was lost in the '50s; it was lost in the '80s, the Reagan years. It's pretty amazingly cyclical." Neither Scott nor Helen lived long enough to see this current back-to-the-land movement. Scott died in 1983, shortly after his hundredth birthday, and Helen died in 1995, at age ninety-one. But both believed in reincarnation, so I like to think perhaps they're nearby, eating Helen's infamous carrot croakers, pulling errant weeds, and relishing this long-awaited moment.

6

FOOD FIGHTS

THANKSGIVING IS THE ONE holiday when my birth family tries very hard to be together—a challenge as we're scattered from New England to the mid-Atlantic to California to Ireland. We're unapologetically guilty of what Helen Nearing called "glutting." My sister and I mostly take turns hosting. She and her children are vegetarian, which turns out to be surprisingly easy to accommodate on turkey day. But the last time Rob and I hosted, my middle brother announced—scant hours before dinner—that he had a special diet, too; he'd given up vegetables. We thought he was goading my sister, but he meant it. My brother Brian's young daughter was eating only white foods. My youngest brother has Crohn's disease and had started avoiding gluten to better manage his symptoms. My dad had been sick again and was on a low-fiber diet. I had long since given up almost all industrial food, ate mostly local. For her part, my mom just wanted a traditional turkey dinner with stuffing and gravy and mashed potatoes and pies, without all this nonsense about special diets.

I'd worked out a cooking schedule that began on Wednesday with down-to-the-minute details of when to do everything from chopping celery to setting the table, in the hope that we could really have everything ready at the same time. Not that we could serve it family-style. Instead, we laid everything out buffet-style, each dish accompanied by a card. MUSHROOM GRAVY—VEG, G FREE; STUFFING—VEG, NO NUTS; STUFFING—NON-VEG, W/NUTS; HEIRLOOM BEANS—G FREE, VEG, FROM OUR GARDEN. A provenance for each dish, half of them variations on one another. Even the mashed potatoes were done two ways: one with cream and butter, one dairy-free for the lactose-intolerant or -avoidant.

Afterward, walking off the main course so we could come back for pie, seventeen of us in a straggling queue along the sandy edge of our quiet street, my mom pressed my forearm gently to get me to slow down.

"That was nice of you," she whispered. "But you don't have to do it, you know."

"What?"

"All those different foods. People can just pick and choose."

"Yeah, but I wouldn't feel like a good host."

"But you cooked for two days for a meal that lasted an hour." My mom shook her head. "Remember what Grandpa Kelley used to say?"

Of course I did. *Take what you're given or go without.*

And that worked well for him, especially during the Depression and World War II. Maybe it could work now if we all had the same food politics (or digestive woes). But we don't. And in that, we're far from alone. My sibs and I grew up eating a pretty conventional American diet, circa the 1960s and seventies, but we've gone our separate ways. Our food preferences and politics have us eating different meals at the same table, at odds with each other, but in diverse good company nationwide. Sensible as the various diets may be (well, not the one without vegetables), they generate conflict in the kitchen and the culture. No one of them is clearly best for meeting the body's caloric and nutrient needs, or for highlighting the social and gustatory aspects of eating. Nor is one of them geographically or culturally specific to my family and our forebears, which is largely how diets used to be determined.

Many contemporary eaters base their food choices on politics and ethics. The variety of resulting frameworks can seem like tacit (or not-so-tacit) indictments of alternative foodways. Is being vegan the only ethical option? Is eating meat to be polite good or bad? Is dairy inherently cruel? How in the world did we get from *take what you're given or go without* to our current array of diets shaped by diverse ethics, calorie counts, ease, disease, and fears? And how, after almost a century of eating otherwise, did it happen so quickly?

In many ways, we have the hippies who went back to the land in the late 1960s and the seventies to thank. Their creation of an alternative food system was intrinsic to the emergence of the current Good Food Movement, as important to it as the contributions of chefs such as Julia Child and James Beard. By the 1960s, "American food" was firmly entrenched, and was one of the many things the seventies-era back-to-the-landers did their best to leave far behind. In its place, they championed organic and unprocessed foods. They also helped successive generations to relearn how and why to grow foods organically and helped establish larger markets not only for those organic goods, but also for local foods, vegetarian fare, and assuredly pure foods.

In making secular ethics important to their dietary choices, the hippies were continuing a practice of their predecessors. For the Fruitlanders, in the 1840s, pursuing a right diet was explicitly an ethical decision; it provided a way to be at one with nature and to protest slavery. For the Brook Farmers, growing food increased their self-sufficiency and offered opportunities for the literary types to do some physical tasks, furthering the overall goal of intermingling mental and manual labor. Thoreau based his diet on simplifying and on sociality; although mostly a vegetarian, he ate what he was offered in other people's homes. For the folks growing their own at Arden, a primarily vegetarian diet enabled them to be producers and to live outside an increasingly inegalitarian, consumerist culture; likewise, the Nearings were able to avoid capitalism and consumerism.

Before we look at the efforts and food decisions the hippie back-to-the-landers pursued, it's helpful to have a fuller sense of the foodways they spurned. To do that, we need to pick up the story of America's changing agriculture and food culture where we left off, at the Great Depression, when the Nearings left New York and moved to Vermont.

THE GREAT DEPRESSION

AT THE BEGINNING OF the Great Depression, Americans addressed poverty and need as they had in earlier times—through local churches, poor farms, and charitable organizations. Bread lines met a small portion of the need, particularly in urban areas. But as hunger became more widespread, governments were forced to step in. School lunch programs, until then mostly ad hoc efforts, became more formal. Knowing the lunches were often children's only meal of the day, cities and states offered guidelines and dietary recommendations.

President Hoover was reluctant to acknowledge how dire the situation was. Finally, in 1932, he conceded that national efforts were necessary. Since 1929, the Federal Farm Board had been buying millions of bushels of wheat from farmers to stabilize the grain's price; Hoover gave the wheat to the Red Cross, which arranged for it to be milled, bleached, and distributed. Much of it went directly to schools for their lunch programs. Some went to welfare agencies to distribute to families. "They were surprised," as Jane Ziegelman and Andrew Coe recount in their culinary history, "to discover that large numbers of American homemakers had forgotten how to bake." The Red Cross had to set up programs to teach them how.[1]

By 1933, as the lack of food was precipitating violence, the Secretary of Agriculture, Henry Wallace, asked President Roosevelt to grant him emergency powers. Roosevelt accorded him the "power to control production, buy up surplus commodities, regulate markets and production, and levy taxes to pay for it all"; Congress affirmed the particulars by passing the Agricultural Adjustment Act. This act required farmers to limit their production of seven basic commodities—corn, wheat, tobacco, hogs, rice, cotton, and dairy. Limiting supply, it was hoped, would ensure that farmers got more money per unit of their goods and simultaneously reduce surpluses. It sounded like a win for all. And it did much good. But it didn't do so across the board; 98 percent of the support went to white farmers, undercutting the gains African American farm owners had made in the early decades of the century.[2]

With those who farmed most of the commodities, the USDA was able to negotiate deals readily. But the Agricultural Adjustment Act came into effect in May—after cotton farmers had already planted for the year. By summer's end, the USDA had finally reached deals with many of them; they agreed to plow under 10.4 million acres of cotton, almost a quarter of their crop, in return for one hundred million dollars. Wallace also negotiated a deal with hog farmers to slaughter some six million piglets, thereby reducing the amount of corn the pigs would need to eat to grow to market size and limiting the amount of pork that would enter the market. The repercussions of these two decisions still ripple through American agriculture.

The subsidies to cotton farmers were meant to be split with tenant farmers and sharecroppers who worked in their fields. But after some politicking, the checks were sent to the farm owners. Many kept all the money, which left farm laborers without any prospect of income for an entire year. Almost a million African American farmworkers were forced off the land. The destruction of the piglets outraged the hungry public. Facing an uproar, Wallace quickly established the Federal Surplus Relief Corporation, to distribute food and items such as lard and soap made with animal products. "Not many people realized how radical it was," Wallace said, "this idea of having the Government buy from those who had too much, in order to give to those who had too little."[3]

For the short term, Wallace's redistribution plan provided an elegant solution to both urban and rural woes. The government bought surplus crops, aiding farmers; it gave those crops to schools for expanded lunch programs staffed by thousands of newly hired workers through the Works Progress Administration. The workers could then afford to buy food, and the promise of lunch ensured that children who attended school had at least one hot meal every weekday. But, as Wallace himself pointed out, plowing under cotton and killing piglets weren't "acts of idealism in any sane society"; rather "they were emergency acts." And in the long run, at least three aspects of these efforts have served the nation poorly.[4]

First, the programs benefited only farm owners who produced certain commodities, almost all of whom were white men with large farms. Farm workers didn't benefit, nor did very many farmers of color or farmers who grew specialty crops.

Second, schools ended up homogenizing students' diets even more than they already were. Early in the century, schools in large urban areas began offering lunch programs to both nourish students and Americanize their diets. As the food historian Harvey Levenstein explained, it didn't matter whether the students liked the food; they were "learning an important lesson: it was the food in their homes, not that on the steam tables, which was out of the mainstream, and that to enter that stream they would ultimately have to learn to appreciate its food."[5]

During the Depression, lunch menus were determined by what the government could buy in quantity, so meals were designed not with flavor, or regional character, or ethnic or traditional preferences in mind. Consider this weekly menu, which was in the New York City school system's rotation:

Monday:	cocoa, tomato puree, succotash, cheese sandwich, fresh or stewed fruit.
Tuesday:	pea soup (without milk), Italian spaghetti with onion and tomato sauce, white rolls with butter, chocolate pudding served with milk.
Wednesday:	tomato macaroni cup (thickened with farina), baked salmon or creamed salmon, buttered graham roll, rich custard pudding.
Thursday:	vegetable cup with rice, creamed lima beans, whole wheat bread and butter sandwich, stewed or fresh fruit or fruit gelatin or fruit salad.
Friday:	lima bean and barley soup, jam or fish sandwich on whole wheat bread, creamed carrots with peas or creamed cabbage or mashed turnips, vanilla cornstarch pudding with chocolate sauce.[6]

Nutritionally sound by the standards of the day, the meals rely heavily on lima beans, a vegetable many children loathe. The required foods "introduced a common culinary language," emphasizing "mild, 'easily digested' fare," but even many hungry children didn't want to eat them. Getting students to try unfamiliar foods "required gentle indoctrination on the part of the lunchroom teachers, a task they were fully prepared to take on. Carefully recording each child's nationality, teachers tracked which foods were willingly consumed and which rejected." If children refused to eat certain items, they were coaxed to try them. If they continued to refuse, parents or older siblings were called in to insist. The result of this behavioral conditioning? A generation of immigrant children who ate what the New York City Board of Education described as "the ordinary American dietary."[7]

And third, farmers came to depend on a guaranteed outlet for commodity surpluses. Knowing the government would help them deal with surpluses gave farmers little incentive to let land rest. The overproduction of dairy, corn, and soybeans grew so acute that many small and mid-sized farms in those sectors were forced out. Ironically, the federal government simultaneously urged destitute urbanites to homestead so they could feed themselves while pursuing policies that disproportionately supported large-scale farms and caused small farms to fail. The back-to-the-landers who fared best in this environment were those who sought to be truly self-sufficient, tending gardens and livestock for themselves, not for the marketplace.

EATING DURING WORLD WAR II

BY 1942, THE FEDERAL government was again sending all it could overseas to feed military troops and provide food aid to allies, but school lunch programs continued because Congress authorized funding for them. The anthropologist Margaret Mead was the executive secretary for the Committee on Food Habits, in charge of establishing a national policy for school lunch programs to make sure the meals helped build "a unified national identity." Wary of spicy foods, which were thought to be unhealthy, she proposed "low toned foods" that are "fairly innocuous and [have] low emotional value." Lunch program directors obliged. In Chicago, with a student population including the children of Polish, Lithuanian, and Mexican stockyard and meatpacking workers, the director of the school lunch program proudly noted that the multiethnic student body "actually ate democracy."[8]

As the war continued and rationing increased, school and community gardens helped expand food availability. An elementary school teacher in Detroit wrote, "During the spring and summer of 1943, Victory Gardening, next to the war, was perhaps the most universally discussed topic," as students in grades four through eight learned to assess soil health, pot up seedlings, tend to the young plants, weed, and harvest, and then to preserve foods. In Chicago, schools partnered with city parks, and Marshall Field & Company provided seeds for beets, radishes, lettuce, chard, kohlrabi, beans, radishes, and flowers to more than fourteen thousand young student-gardeners. Nationwide, almost one in three urban families tended a Victory Garden; on average, they provided 2.8 pounds of vegetables per person per week in season.[9]

Despite the best efforts of these many gardeners, people were hungry. When I asked my dad what he remembered about foods then, he said he didn't remember much about them. Instead, what he vividly recalled was how difficult it had been to get enough: "I remember my mother taking me and your aunt Lois—you need to remember I was born in 1937 and Lois was born in 1938, so we were very small—taking us by the hand, and we'd walk to the streetcar, and take the streetcar to Blair's Food Market. It had the best selection. But even there, there wasn't much." He paused, lost in the recollection. "My mother would give me some tokens and some money," he went on, "and stand me in one line to buy eggs, and she'd give Lois some more tokens and stand her in a different line. My mother would stand in another line, too, and keep an eye on us, make sure we stayed in our lines, and we would get whatever we could get. And some things, you just couldn't get, like butter."

My dad was five and Lois four when they began standing in line to help get provisions for the family, and like the grammar school students around the nation, they soon also helped weed the family's Victory Garden. My mother, who grew up in a small town, also recalls standing in ration lines and still refers to that period as "the year of the cabbages," the staple her uncle grew.

THE NEW FOOD ORDER

THREE OTHER CHANGES TO America's food system tied to World War II were even more important than Victory Gardens and other efforts to fight hunger. Taken together, they transformed American food. Those changes are the consolidation of the role of the large supermarket, the explosive growth of processed and fast food, and the new scale and style of farming due to industrial fertilizers.

When the United States entered World War II, 8,175 supermarkets peppered the country. Despite food rationing, the number of supermarkets almost doubled to fourteen thousand by 1950. And by 1960, the nation's thirty-three thousand chain supermarkets accounted for 70 percent of sales for at-home foods. They edged out not only neighborhood "mom-and-pop" grocery shops, but also most of the local farmers, bakers, greengrocers, butchers, and fishmongers. Their ability to offer lower prices fueled this meteoric rise. At small stores, a customer gave her shopping lists to a clerk who gathered all the items and delivered them—for free—to the customer's home. Supermarkets, in contrast, were self-serve and did not provide delivery, so they needed far fewer clerks. The lower labor costs were passed on as savings to customers, which increased customer loyalty, giving the supermarkets more leverage to negotiate lower prices from wholesalers.[10]

The second war-induced change was a surge in the variety of processed foods and kitchen gadgets available. In a 1957 *Washington Post* article entitled "Modern Kitchen Miracles," the reporter raved about the new time-saving foods, appliances, and other kitchen tools the up-to-date housewife could enjoy, such as "frozen orange juice, instant cocoa, cookie mixes, 'brown and serve' rolls, dehydrated soups in little plastic bags, freezers, blenders, cakes ready to bake in throw-away foil pans, no-rinse detergents, fast oven grease removers, pancake batter in paper containers, pre-stuffed turkeys, plastic wrap, ready-made dishes, 'TV dinners.'" I was startled to discover that my kitchen, the one I thought old-fashioned in its paucity of processed goods, houses several "miracles"—including a freezer, a blender, and two rolls of plastic wrap.[11]

The *Post* columnist continued with, it seems, even more zeal about other products that were "in the works," like "automatically irradiated foods that last indefinitely without refrigeration, and the 'pouch' dinners—whole dehydrated meals in plastic bags that cook instantly when you pour hot water into the bag. Instant spaghetti with instant sauce, bread that stays soft three months, canned pre-fried bacon that heats without splatter."

Exactly how did Americans get so lucky? Why, thanks to the work done to feed soldiers in the field, especially that of the Quartermaster Food and Container Institute for the Armed Forces, whose job is "improving military meals—but women owe it a huge vote of thanks: It put cake mixes, dehydrated soups and boneless meats in your kitchen . . . When the war was over, powdered eggs, dehydrated potatoes, etc., were grabbed by civilian industry . . . These solutions became today's convenience foods." I assumed some of these miracle foods were surely dreams deferred, but even pre-fried bacon, the one I most doubted I'd find, is readily available.

Indeed, many products popular today were introduced during the 1950s—Kellogg's Sugar Pops and Sugar Frosted Flakes, Tropicana juices, Swanson's TV dinners, Stouffer's frozen meals, Mrs. Paul's Fish Sticks, Pillsbury's refrigerated cookie dough, Duncan Hines's cake mixes, Star-Kist canned tuna, Cheez Whiz, Sweet'N Low sugarless sweetener, Carnation Nonfat dry milk, and Lipton Instant Tea among them. Complementing the promise of speedy home cooking were speedy restaurants: the first McDonald's opened in 1948, Dunkin' Donuts in 1950, Taco Bell and Jack-in-the-Box in 1951, Denny's in 1953, Burger King in 1954, Kentucky Fried Chicken in 1955, and Pizza Hut in 1958.

The core of contemporary food culture was firmly in place by mid-century. Financially more secure than they'd been in years, Americans bought refrigerator-freezer combos and deep freezes and filled them with purchased frozen foods. Like the mania for canned foods half a century earlier, Americans in the early 1950s craved frozen everything. Sales of frozen items rose from $496 million in 1949 to $2 billion in 1956 as Americans spurned fresh foods when frozen versions were available.[12]

Part of the appeal of frozen foods was ease. They relieved cooks of tasks like shelling peas, de-worming broccoli, and shucking corn. But they were also part of a broader shift away from cooking from scratch. Name-brand products and mixes appealed to cooks striving to be up-to-date. The historian Jessamyn Neuhaus reviewed more than a hundred popular cookbooks from the 1950s and found the recipes routinely included processed and packaged foods specified by brand. Even soup recipes were usually instructions for doctoring a canned soup. To be sure, some magazines of the era, *Living, Life,* and *Ebony* among them, tried to appeal to women's "gracious living" aspirations by waking up their inner gastronome. But like most aspirational writing, they did more to capitalize on the gap between reality and aspiration than to close it.[13]

The third major change wrought by the war, the introduction of chemical fertilizers, altered what farmers grew, how much they grew, and how they grew it. Before World War II, food farmers used few synthetic fertilizers, relying instead on compost, manure, and other natural sources of nutrients. In the wake of the war, munitions plants that had been built to supply the government with ingredients for explosives were repurposed. They began producing nitrogen-rich chemical fertilizers, making available a virtually endless supply of one of the key nutrients plants require.

Much as agricultural capitalism changed farming in Thoreau's era, much as improved tools and irrigation techniques post-Civil War pushed farmers to expand, access to chemical fertilizers led to another massive change in agriculture in the mid-twentieth century. In each transition, small-scale farming became more financially tenuous. The fifty- or hundred-acre farms that had supported a family suddenly couldn't; once again, farmers had to expand to recoup fixed costs—or they had to quit. Both happened: "Between 1950 and 1975, the number of farms in the country declined by half, as did the number of people on farms. And the average size of farms nearly doubled, from 216 acres in 1950 to 416 acres in 1974." Half as many farmers were tending twice as much land, and the majority were planting just one or two crops rather than a diverse array. This was precisely what the USDA had long urged: It maintained production could be improved "by winnowing out the most vulnerable farmers, consolidating landholdings, and subsidizing mechanization and chemical insecticides."[14]

In California at the end of World War II, landholdings were extremely consolidated; just seventy-eight growers owned roughly six million acres of farmland. With so few companies determining how so much land was tended, "corporations wielded important economic and political power and influenced marketing, labor, and water issues in California and the Far West." To describe what was happening on those mega-farms, which relied on new technologies and a large but low-skilled workforce, the state's Director of Immigration and Housing coined the term "factory farming." The power of mega-farms continues to be evident in the ways immigrant labor is incorporated into producing and processing food.[15]

ORGANICS AND FOOD ACTIVISM

LIKE THOREAU BEMOANING THE rise of agricultural capitalism, able to see some of its limitations in the moment, a few people recognized early on that chemical-industrial agriculture wasn't an unalloyed good. The English botanist Sir Albert Howard was among them; his *aha!* moment led him to become an organic-farming pioneer. Sent to India to teach farmers in the then British colony about up-to-date farm techniques, Sir Albert learned from them instead. Their reliance on compost, rather than synthetic chemical fertilizers, reaped healthier soil, plants with greater resistance to pests, and foods with better flavor. By 1928, Sir Albert was convinced compost was preferable to chemical inputs. He continued to study its efficacy, and in 1940 offered detailed accounts of the Indian farmers' practice in *An Agricultural Testament*.

A few years later, the American publisher J. I. Rodale discovered Sir Albert's work, and was so affected by it that he quickly wrote and published *Pay Dirt: Farming and Gardening with Compost*, the first of his many texts advocating sustainable agriculture and organic methods. Like Sir Albert, he emphasized that soil was a living entity, and simply adding nitrogen, potassium, and phosphorus didn't generate the vast microbiome found in healthy soil. Instead, he said, farmers should "observe the Law of Return, restoring to the soil all plant residues that came from it" and incorporating animal manures that are "well-rotted." When done properly, organic growing practices do more than simply return soil to its pre-use state. They create a virtuous cycle among people, plants, and soil. People improve the soil fertility; fertile soil enhances the plants' growth and nutrient density; nutritious plants improve the physical well-being of those who eat them.[16]

In 1949, Aldo Leopold made a related observation in a more philosophical tone. He said ethics had not yet been extended to the human relationship with the land, but they ought to be. "An ethic," he wrote, "may be regarded as a mode of guidance for meeting ecological situations so new or intricate, or involving such deferred reactions, that the path of social expediency is not discernible to the average individual." A land ethic, he said, is not fundamentally different from other ethics; it "simply enlarges the boundaries of the community to include soils, waters, plants, and animals, or collectively: the land." Such an ethic is what we see in the different generations of back-to-the-landers—they all express a sense of reciprocity with their land, crops, animals, and communities.[17]

In 1971, after his father's death, J. I. Rodale's son, Robert, took the helm of Rodale Press. Among the first books to come out under his leadership was *The Basic Book of Organically Grown Food*, which emphasizes the social and cultural benefits of organics, an expansion of the "boundaries of the community" impacted by organic methods. Its authors note that "by wanting to buy organically grown foods raised by a family farmer who is not supposed to be able to make a living on the land, you're helping to reverse a trend that has driven people off the land and made farming an old man's profession. By buying organic foods at a mama-and-papa neighborhood store that is not supposed to be able to compete with supermarket chains, you are helping to change the make-up of America."[18]

This expanded framing of the benefits of organic farming and organic food did not mention race, but some of those who most quickly embraced organic methods were young adults who had already sensed the political potency of food while protesting race-based injustices. As Warren Belasco points out in *Appetite for Change: Why the Counterculture Took on the Food Industry*, both civil rights advocates and United Farm Workers allies employed food in their protests.

Civil rights activists who organized sit-ins at segregated lunch counters and restaurants in the early 1960s were claiming for themselves the ability to decide what, where, and with whom to eat. The success of those campaigns in desegregating lunch counters in many southern cities also helped desegregate other public spaces, including hotels, theaters, and restaurants. The sit-ins were led predominantly by African American students, but white student activists participated and often brought the skills they honed as protestors back to their colleges. There, they were able to rally their peers to collective action, initiating, among other efforts, boycotts of grapes and lettuce in solidarity with the United Farm Workers.

The impact of the grape and the lettuce strikes in California fields stretched across the nation as labor union members, the Black Panther Party, indignant college students, and others boycotted the produce. Workers on railroads delayed shipments, so grapes rotted before reaching their destination. The BPP joined the UFW in boycotting Safeway grocery stores, using their platform—their weekly paper—to disseminate information about the strikes and boycotts. The BPP even banned consumption of its official drink, "Bitter Dog," because it contained California red wine. On college campuses, boycotts were lengthy and often helpful. The inevitable "Let Us Boycott Lettuce" topped an open letter to the director of food services in Vassar College's newspaper on November 12, 1972. And the article "Farm Workers Press Lettuce Boycott" in the *Harvard Crimson* begins, "Picketing of several Harvard dining halls to protest the University's continued purchasing of non-union lettuce has generally been successful." Eating together—or not eating together—became a more common mode of social protest.[19]

Awareness of the need for better working conditions for food laborers did not immediately translate to individuals buying organically grown foods, much less to growing their own. Organically grown produce often looked less appealing than its industrial counterparts. More than a decade later, when I was buying food at Bloomingfoods Co-op, the vegetables were frequently misshapen or pest-scarred. Compared to the slick appearances of industrial crops, selected for transportability and shelf appeal, bedraggled carrots, pocked apples, and scabbed potatoes certainly didn't look like the healthier option. Happily, a gnarled appearance was anything but a turnoff to many of the folks who would embrace organics and go back to the land in the late 1960s and into the seventies.

DIRTY HIPPIES AND DIGGERS

WHEN I WAS STILL in grade school, the same time many hippies were heading back to the land, teachers offered us the pithy vocational advice that there were "two kinds of jobs, the ones you shower before and the ones you shower after." We knew which sort we were supposed to want. Americans have long elevated cleanliness to almost sacramental status— the outward sign of economic success, middle-class respectability, and moral integrity. It's hardly surprising, then, that the long-haired, full-bearded folk who flouted the social standards of being "clean-cut" and "clean-shaven" were often called not simply "hippies," but "dirty hippies."

Their purported lack of interest in personal hygiene was often complemented by a deep concern about other kinds of clean. To be sure, many hippies were simply dropping out, doing their own thing. But many others were becoming the vanguard of the environmental movement, working to ensure clean food, clean water, clean air, and clean land. Those who wanted a more wholesome environment were perfectly happy to eat crooked organic carrots or tired-looking dandelion greens they'd foraged or cheese with a moldy corner they had to trim off.

Even before the full-fledged turn to the land began, even before the embrace of organic methods, some hippies in San Francisco's Haight-Ashbury were laying the foundation for a radical rethinking of food and food systems. By the summer of 1966, Haight-Ashbury had assumed much of the character for which it became infamous the next year. Lots of hippies had moved in; pot and LSD were blowing people's minds; the groovy vibe was imbued with "a spirit of generosity," in the words of one of the folks I interviewed. The neighborhood wasn't yet full of runaways, wasn't yet a gawkers destination, wasn't yet dominated by heroin. Sometime that summer, Billy Murcott, a young New Yorker, moved there to reconnect with his boyhood friend Emmett Grogan. A beautiful, charismatic anarchist, Grogan was working on the fringes of the San Francisco Mime Troupe. Through the group Murcott also met Peter Berg and Peter Cohon (a.k.a. Peter Coyote).

Berg and some other members of the troupe were reimagining the way they might work—elaborating on the guerrilla theater they'd been doing under the direction of the founder, Ronald Davis. Berg envisioned using "theater to cause something to happen and what happened would be an issue in itself, and it was important for people in the audience not to be told it was theater. Not to sneak it on them, but to have them perceive it as reality, and have them relate to that reality, to create real change." Berg thought performing as "life-actors" would have more impact. But Davis wasn't interested, so Berg and Grogan resigned from the troupe to pursue their ideas elsewhere.[20]

At roughly the same time, Billy Murcott was reading about anarchy in England and learned about the Diggers, seventeenth-century rebels who tried to create small agrarian, egalitarian communities. The Diggers were reacting to more of the commons being enclosed. In the early 1500s, when Sir Thomas More wrote *Utopia*, enclosure was just starting to be a concern; by the time the Diggers protested, in 1649, poor people had far less access to land, so even fewer chances to grow or gather food. The Diggers thought land should be turned over to the people so they could grow food and live in an egalitarian way; they hoped such communities would inspire others. (How similar this sounds not only to More's *Utopia* farmers, but also to the Fruitlanders' belief that humans can't own land, and to folks inspired by Henry George's single-tax model.) In generative protest, the Diggers squatted on enclosed plots, planted vegetables, and invited others to join them.

Loving the original Diggers' idea of simultaneously claiming freedom and aiding those in need through direct action, Grogan, Murcott, Berg, and Cohon promptly dubbed themselves the Diggers. Murcott, whom Peter Coyote called "the unsung genius of the Diggers," was most responsible for articulating the anarchic group's principles and composing many of its broadsides. The first anonymous broadside, "Let Me Live in a World Pure," chided not only establishment sell-outs, but also those in the counterculture who weren't fully living up to their principles, and asked questions like "When will the JEFFERSON AIRPLANE and all ROCK-GROUPS quit trying to make it and LOVE?" During their time in Haight-Ashbury, the Diggers posted provocative broadsides, staged all manner of street theater, and carried out a crazy array of pranks and exploits. But by far their most enduring contribution to the counterculture was one of the earliest.[21]

On September 27, 1966, a white police officer shot an unarmed sixteen-year-old African American named Matthew "Peanut" Johnson. He'd been joyriding with friends; when police stopped them, the teens fled. An officer shot Johnson in the back, killing him. Hours later, rioting and looting began and quickly spread from Hunters Point to the Fillmore District, closer to Haight-Ashbury. The National Guard was called in. San Francisco imposed a curfew and did so again the next night, after more rioting broke out. Police and National Guard were authorized to shoot to kill.

Emmett Grogan and Billy Murcott were on Fillmore Street eating barbecue when the area was closed off by the police. Grogan recalled that they "stood outside the eatery, watching the surreality of the paramilitary operation unfold on the sidewalks where children had been playing only moments before. They got caught up in a crowd of black people who were trying to get back to their homes, and it felt strange being white but nobody said anything." But then "some kids started throwing bricks from the vacant lot at an official car that was passing, and suddenly someone hollered, '*Lock 'n Load!*' Fifty cops immediately dropped to their knees and jacked shells containing pea-sized charges of buckshot into the chambers of their riot guns."[22]

Grogan wrote that his first impulse was to climb onto a roof and throw Molotov cocktails at the police and soldiers, but he quickly realized doing so "would touch off a murderous, repressive onslaught by the soldiers, in which black people would suffer a far more devastating and wholesale repression than they did already." Rather than add to the strife, the pair vowed to come up with "some sort of score" that would lead to a heightened "collective social consciousness and community action."[23]

A day or two later, they posted mimeographed announcements throughout the neighborhood:

FREE FOOD GOOD HOT STEW

RIPE TOMATOES FRESH FRUIT

BRING A BOWL AND SPOON TO

THE PANHANDLE AT ASHBURY STREET

4 PM 4 PM 4 PM 4 PM

FREE FOOD EVERYDAY FREE FOOD

IT'S FREE BECAUSE IT'S YOURS!

the diggers.

That switch in gears—putting aside Molotov cocktails and making free food for anyone who wanted it—was genius. So genius that despite having taken their name from the original Diggers, some of the San Francisco Diggers didn't get it at first. In his memoir, Peter Coyote described the first Digger Free Food he attended with Emmett Grogan, watching a line of people being served stew and loaves of bread that looked like mushrooms because they "had been baked in a one-pound coffee can and had expanded over the top to form a cap." To get the meal, each person had to step through "a large square constructed from six-foot-long bright yellow two-by-fours" that the Diggers called the Free Frame of Reference. Emmett Grogan asked Peter Coyote if he wanted to eat, and he said, "No, I'll leave it for the people who need it." At that moment, Coyote wrote, Grogan "looked at me sharply. 'That's not the point,' he said . . . The point was to do something *you* wanted to do, for your own reasons. If you wanted to live in a world with free food, then *create* it and *participate* in it. Feeding people was not an act of charity but an act of responsibility to a personal vision."[24]

Firm in their belief that food should not be corporatized because "every brother should have what he needs to do his thing," the Diggers created an alternative food network. Digger volunteers "hit every available source of free food—produce markets, farmers' markets, meat packing plants, farms, dairies, sheep and cattle ranches, agricultural colleges, and giant institutions (for the uneaten vats of food)." Then they filled up "their trucks with the surplus by begging, borrowing, stealing, forming liaisons and communications with delivery drivers for the left-overs from their routes." A morning shift gathered the food and an afternoon shift delivered it to "the list of Free Families and the poor peoples of the ghettoes." With every free meal, with every volunteer food run, the Diggers performed their utopia.[25]

Like those who lived in experimental communities in the 1840s, or the Ardenites staging community in their fêtes and celebrations, or Helen Nearing enacting the good life so others saw what it could look like, the Diggers acted a new reality into existence. That reality was rooted in the desire to be free: Food had to be free because it was a necessity, and people needed to exit capitalism because it curtailed freedom.

In "Trip Without a Ticket," the Diggers wrote that the riches of America and the power of computers "render the principles of wage-labor obsolete . . . We are being freed from mechanistic consciousness. We could evacuate the factories, turn them over to androids, clean up our pollution." Although the Diggers' idiom is a far cry from Scott Nearing's flat prose, their observations are astonishing in their similarity: "Our conflict begins with salaries and prices. The trip has been paid for at an incredible price in death, slavery, psychosis." In another manifesto, "Men Are Out of Touch with the Earth," they describe capitalism's high cost for individuals in ways that evoke Thoreau's early critiques of wage labor: "We have plunged into a luxury of impoverishment. Money has reckoned the soul of America. The state imposes the elements of survival: frozen foods. Each element appears to be a liberation and turns to servitude."[26]

In the months after the Diggers began the daily Free Feed, hippies continued coming to Haight-Ashbury. Many of the young runaways embraced the "Digger thing" of giving items away, most often flowers, fruit, candy, and hugs. As Alex Forman wrote at the time, "The Diggers in a sense became a new morality, the opposite of industrial capitalism's grab-bag marketplace morality." Unfortunately, as he pointed out, they "began to be labelled by some as a 'community service.' It was at this point that an inevitable split occurred, for the Diggers did not want to be a community service; they wanted the community itself to be based on the new morality."[27]

The community wasn't ready. As more and more young people arrived, conditions deteriorated. David Simpson, a former Digger, said the group created "a new, free society in the shell of the old," but Emmett Grogan was less sanguine, even when the free food in the Haight was being supplemented by Free Stores, a Free Clinic, Free Festivals, and lots of free music. The community was growing less coherent as more runaways and addicts—and the people who prey on them—showed up. Whatever capacity the community had for transformation in 1966, it was gone by the end of the Summer of Love, the following year.[28]

Gone among the mostly white hippies but not from the Black Panther Party in nearby Oakland. The BPP expanded significantly at the end of the sixties, and in January of 1969, it started the Free Breakfast for Children Program, the first of its "survival programs." Like the Diggers' assorted free enterprises, the survival programs were intended to provide free goods and services to help change people's consciousnesses and to promote their capacity to become free. Like ingredients for the Digger Feeds, food for the Free Breakfasts was donated—sometimes out of generosity, sometimes after some arm-twisting. In pressuring businesses reluctant to help feed children, the BPP highlighted the link connecting capitalism, economic inequality, and personal experiences of hunger, which both shamed hesitant store owners and proved amazingly effective in catalyzing political consciousness.

When Emmett Grogan learned that Oakland's chapter of the Black Panther Party was planning a Free Breakfast Program, he immediately headed over with food. According to the Panthers' chief of staff at the time, Grogan was the first person to donate, and he continued to do so regularly. The Free Breakfast Program grew rapidly, feeding eleven children the first day and one hundred and thirty-five by the end of that first week. Soon, it was replicated in forty-five city neighborhoods across the nation, feeding some fifty thousand children each day. Although the BPP's Free Breakfast Program lasted less than a decade, it helped reframe hunger and the associated lack of freedom as the result of structural failings. As state and federal governments stepped in to feed more children, they amplified the nutritional impact made by the Diggers and Panthers. In 1975, Congress expanded the School Breakfast Program to all public schools.[29]

The Black Panther Party remained primarily urban and focused primarily on urban issues, but many white hippies left cities, decamping to the countryside. Some who'd been part of the Haight-Ashbury scene went to Lou Gottlieb's Morning Star Ranch, in Sonoma; some formed the commune Black Bear Ranch in Siskiyou County; others headed to New Mexico—so many, in fact, that Taos was nicknamed "Haight-Ashbury East."

The Diggers played a critical role in promoting the back-to-the-land culture. Hippies in New England often credit the Nearings with inspiring them, but those in Northern California, the

Pacific Northwest, and New Mexico are far more likely to mention the Diggers, or the hippie scene in the Haight more generally. Their anticapitalist stance and yearning for freedom—traits they had in common with Thoreau, the Brook Farmers, the Fruitlanders, the Ardenites, and the Nearings—pervaded their back-to-the-land ethos. In fact, like Alcott and Lane before him, Lou Gottlieb believed so strongly that private property was anathema that he transferred the deed for Morningstar Ranch to God. And like the Ardenites, many saw their rural communities as way stations for burned-out urban activists. The Tick Creek Tribe wrote, "Living outside the pig economy, gaining intimate, sure control over the mode & quality of our lives—these are revolutionary activities. They extend the outlaw area which supports and serves and—when necessary—hides us."[30]

As they moved back to the land, communards and householders retained their awareness of the revolutionary power of food. In kitchens across the nation, homemade whole wheat bread like the loaves the Diggers distributed became a staple. An emphasis on affordable, nutritious food cemented in the countercultural consciousness a sense that what people eat and how they get their food are politically charged issues, central to creating the kind of society one wants. They made clear that those choices could support the existing system or support alternatives to it. And they chose change.

7

HIPPIE FOOD

I N THE SPRING OF 2017, the *New York Times* ran an article headlined "The Hippies Have Won," which opened this way: "It's Moosewood's world. We're just eating in it." The reporter hailed the mainstreaming of granola, kombucha, miso, soy, and other countercultural staples as proof that we've become groovier eaters. Their place in the American pantry and on the American plate had been secured, this article held, partly by chefs who made those ingredients sing in ways hippie-era communards mostly didn't and partly by industrial food producers cashing in on yearnings for healthier options.[1]

The seeds of "Moosewood's world" originated in Berkeley, and were transported east to Ithaca, New York, by Mollie Katzen. A student at Cornell, Katzen had worked at some of the region's vegetarian restaurants. When the university shut down for the second time in as many years, she headed to Berkeley to finish her degree. To help cover expenses, she took a job at a vegetarian restaurant there, and was amazed. Unlike the beige macrobiotic foods common in New York's vegetarian restaurants (which she called "remorse cuisine"), California's vegetarian fare was colorful, with fresh ingredients, bright sauces, and lots of variety. Returning to Ithaca, she joined a group in the process of planning a restaurant, the Moosewood Collective. As the member with the most restaurant experience, Katzen served as menu maker and head cook for Moosewood Restaurant, which opened in 1973.[2]

A year later, she published a small, spiral-bound cookbook of the restaurant's most popular recipes. It quickly gained a cult following, prompting Katzen to work with a publisher who could distribute a revised version on a national scale. In its first five years, *The Moosewood Cookbook* sold more than a quarter of a million copies, and in the decades since, it's remained hugely popular. According to the *New York Times*, it's one of the top ten best-selling cookbooks ever and it's in the James Beard Foundation's Cookbook Hall of Fame.

I imagine that Mollie Katzen, who called cooking from scratch "a radical necessity," would balk at describing items like, say, Quaker Oats Chewy Chocolate Chip Granola Bars as Moose-woodian. Still, the *Times* article reminds us just how much of what's now common in the food world—from attention to local food to widely available organics to half a supermarket aisle of granola and granola-adjacent products—is indebted to those who went back to the land in the late 1960s and the seventies.[3]

Creating "Moosewood's World" wasn't the main aim of those who went to the land in the 1970s. In *Getting Back Together*, Robert Houriet, himself a hippie back-to-the-lander, wrote that the communards didn't *have* broadly shared aims; "only afterwards was it called a movement. At the outset, it was the gut reaction of a generation." But a movement it was: The most common estimate is that a million young adults went back to the land in communes or as homesteaders. And the many idealists did agree on at least one thing—society was heading in a bad direction and to get on a right path they needed "to go all the way back to the beginning, to the primal source of consciousness, the true basis of culture: the land."[4]

Sometimes deep in the woods or deserts, more often on the outskirts of settled communities, homesteaders and communards tried to create miniature utopias where they could live self-sufficiently, simply, in harmony with nature and one another. Many back-to-the-landers sought to re-create an agrarian, preindustrial lifestyle, much as the Ardenites had in their embrace of the Arts and Crafts movement's imaginary medievalism. Others extolled Native Americans, as Thoreau did. The particulars of their experiments were akin both to one another and to those of their predecessors: They built smaller homes, experimented with alternatives to the nuclear family, sought self-sufficiency, cultivated deeper connections to the natural world, and tried to imbue their lives with more meaning and a sense of freedom.

Some of the communes were quite remote, but more of them were like Thoreau's plot at Walden Pond—within a few miles of a mainstream community. Proximity made it easy for hip folks to travel from commune to commune in search of a perfect fit, as many had in the 1840s. But it also made it easy for curious straights to stop by, along with neighbors, social workers, and sheriffs wondering what these odd new people were doing. Depending on the temperament of the communards and the locals, relationships ran the gamut from cordial to extremely hostile. New Buffalo, outside of Taos, New Mexico, attracted so many people that dozens of other communes were established nearby, leading to deep tensions with the area's established residents. Lou Gottlieb's Morning Star Ranch, in Sonoma County, California, was bulldozed four times by local officials who didn't want an open-to-all-comers commune in their backyard.

Though the Baby Boomer generation touted happenings, spectacles, sit-ins, be-ins, and other forms fusing performance with reality, the utopians among them grew less willing to perform their lives for a curious public than their forebears had been—mostly because it often had bad consequences. When the founders of Oz arrived in Meadville, Pennsylvania, for example, they went to the local newspaper office and sat for an interview, during which they explained they intended to be "a model for the counterculture of the future, when capitalism would be replaced by a brotherhood of love and men would strive to attain cosmic consciousness"—aims a lot like the ones Alcott and Lane described in the letter they sent to newspapers.[5]

But the denizens of Oz hadn't anticipated how much attention they'd get. They became a Sunday-drive destination; townspeople "packed their kids into the family auto and went to gape. The slow procession of cars . . . crept past the farm, churning up a continual dust cloud. Some days more than a thousand cars filed past, so many that state police had to be assigned there to direct traffic and post 'no parking' signs." The Oz family quickly wore out their welcome.[6]

Despite mainstreamers' curiosity, most communes offered little to titillate gawkers. For idealists who wanted their experiments to last, free love and ecstatic tripping had to take a backseat to meeting more basic needs, such as food and shelter. Robert Houriet observed that "a lot of effort, thought and discussion go into the preparation of food, not only because it's a common need, like clothes or housing, but also because food can be a direct vital expression of man's relationship to the whole life cycle." In 1971, intrigued by Houriet's observation, twenty-six-year-old Lucy Horton hitchhiked cross-country, visiting communes and collecting recipes for a cookbook. Spending time at forty-three communes in fifteen states and southern Canada, she confirmed that a new food consciousness was emerging: "[T]he basic idea was a diet based on what's available locally, prepared nutritiously, getting away from meat and fat and sugar." Food—not sex or drugs—shaped community life, with "group cooking the central material fact and preoccupation." Like Bronson Alcott and others before them, the communards believed "ordinary farming" was not the point; instead, food choices provided ways to reengage with the natural world, to mutually nurture self and planet, and to do so without exploiting others.[7]

Given how many people went back to the land in that era, I'm sure plenty of folks weren't terribly concerned about food. I'm in-

terested in the many who were, the folks who integrated their political and philosophical ideals with their culinary habits. Some of
their recipes and the beliefs shaping them have been preserved in
Horton's *Country Commune Cookbook*, as well as in the many other
cookbooks of that era written by collectives and homesteaders.
Darra Goldstein, a food scholar and former editor of *Gastronomica*, considers counterculture cookbooks "so much more than
cookbooks. They were a way of being in the world." That they
were. They reflect the producers' utopian ideas and show how
they enacted them through the ingredients they used, the meals
they produced, the strategies they used to navigate group dynam-
ics about cooking, kitchen work, and hospitality. The cookbooks
offer their makers' experience as guidance to readers interested
in pursuing similar paths. In melding the theoretical with the
concrete, countercultural cookbooks reveal both what those
seventies-era utopians valued and what they sought to avoid.[8]

A COOKBOOK FOR EVERY POLITIC AND PALATE

In the 1967 film *The Graduate*, the main character, Benjamin
Braddock, has recently finished college and a friend of his father
gives him one word of job advice—"Plastics." But Braddock, like
those who were going back to the land when the movie came
out, rejected the plastic world. The back-to-the-landers turned
their backs on a world they saw as televised, corporatized, and
overly mediated. They wanted to feel more, to touch more, to
be immersed in a more natural world. They yearned for something more tangible, palpable, real. When it came to food, the
items they rejected as plastic were wrapped in it, were grown with
chemicals, contained additives with unpronounceable names,
and were always available everywhere. Their alternatives were
natural foods that were grown organically, were as unprocessed
as possible, and were available seasonally, regionally.

In turning away from mainstream food, many back-to-the-landers became vegetarians. Some wanted to eat lower on the food chain for environmental, financial, ethical, or political reasons; others simply liked thumbing their nose at the meat-eating bourgeoisie. In *Diet for a Small Planet*, which came out in 1971, Frances Moore Lappé presented one of the first book-length articulations of the negative environmental impacts of meat production. She pointed out how wasteful it is and argued that changing to a meatless diet would ensure an adequate food supply even if the human population continued to grow. To meet our need for protein, she made a case for increasing soybean production, as soy is an environmentally kinder protein than meat is. Combining an animal-rights case, an environmental case, and a food-supply case for being a vegetarian, *Diet for a Small Planet* struck many responsive chords; it sold two million copies during the seventies and remains important today.[9]

The feminists who established the Bloodroot Collective, in Connecticut, also embraced vegetarianism, doing so as part of their commitment to respect all species and to avoid both capitalist and patriarchal domination. In the opening pages of *The Political Palate*, they express opposition to "the exploitation, domination, and destruction which comes from factory farming and the hunter with the gun." They cooked and ate seasonally in order to be less "disconnected from organic or natural time keeping." And they worked collectively, each doing what she most loved doing—a plan that would have delighted Charles Fourier. Bloodroot, the restaurant they established in 1977, is still open—and still has no cash register, no tipping, no waitstaff, and no bussers because "things are a little different" at "a feminist vegetarian restaurant."[10]

Not everyone who went back to the land became a vegetarian, however. Lucy Horton, who had majored in classical archaeology at Bryn Mawr, paid attention to the differences among the diets of communards in different regions. In New England, she noted, communards were more likely to be omnivores than were those in the West and Southwest. Half the communes she visited in the Northeast had livestock, and their residents "offed their surplus roosters with alacrity."[11]

Although based in California, Maxine Atwater promoted an omnivorous diet. Her *Natural Foods Cookbook*, a little brick of a book printed on crinkly light brown paper, held that carnivores could also pursue an ennobling food politic if they supported organic agriculture. Atwater encouraged readers to "by-pass mass produced supermarket meats packaged in plastic" in favor of that in natural food stores, which "comes from animals raised on organic, natural feed. They have been free to graze and to choose instinctively from the land what is good for their growth." Though concerned about making sure the animals were well treated, she was also motivated by a concern for people, and worried antibiotics and other chemicals the animals had ingested would "pass from the meat into human bodies, possibly causing cancer, sterility and other, not fully known, bodily defects."[12]

Perhaps the most pervasive changes to the American diet came thanks to the back-to-the-landers' and other hippies' interest in macrobiotics. First espoused by George Ohsawa in the 1930s, macrobiotics didn't gain much traction in the United States until after 1966, when Michio and Aveline Kushi cofounded Erewhon, a natural food store that emphasized macrobiotic foods, in Boston. Intriguingly, although Bronson Alcott would have turned up his nose at the macrobiotic emphasis on rice (for the noble reason that, in his place and day, rice was a product of slave labor), he would have liked most of its other dietary rules, which prohibit red meat, sugar, alcohol, dairy, refined flours, yeast breads, and tropical fruits.[13]

Separated as they were by a century and an ocean, Alcott's views were rooted in an ascetic quasi-Christian version of transcendentalism and Ohsawa's in Japan's Zen Buddhism. But the diets they advocated arose from the same basic belief: Both thought eating correctly helped align the individual with the universe, a necessary step in freeing the self from material limitations. Alcott thought people could abstain their way to bodily and spiritual well-being. Ohsawa thought individual illness and unhappiness were due to an imbalance of yin and yang in the body, and that broader, societal woes were due to an imbalance of yin and yang in the world. Pursuing a macrobiotic diet would bring an individual's yin and yang into equilibrium; then the individual could come into balance with the world. Ohsawa also believed eating these foods with wooden utensils from a wooden bowl (as the Nearings did) was optimal for reestablishing the yin-yang equilibrium.

In 1968, when the macrobiotic-diet adherent Kathryn (Ash) Hannaford moved to San Francisco, the only eatery offering such options was Here and Now, in Haight-Ashbury. The first vegetarian restaurant in the city, it was established a year earlier by Allen Noonan Michael, who had also formed the One World Family of the Messiah's World Crusade, a commune whose members worked at Here and Now. Frequenting it, Kathryn spent a lot of time talking to Michael and became interested in the One World Family, especially its members' commitment to equality and total sharing. Over the next few years, she cooked in the family's restaurant, eventually publishing many of its recipes in *Cosmic Cookery*, a cookbook for people who wanted to become vegetarians.

There, she explained that the One World Family practiced vegetarianism because that diet accords with a higher-order, universal harmony that most "earth people" can't recognize because they've been "conditioned by a material science that . . . isn't aware yet of primal nature." Despite earth peoples' blinkered view, the universe won't let humans fall into utter degeneracy; it "steps in—as it is now through the media of the UFO's, its 'Angels' or higher space beings, who are the overseers of this planet—when the time comes for earth people to be brought out of their time-lag and into the fulfillment of their New Age." For the One World Family, as for the Fruitlanders, Fourierists, and others, diet could "bring about a Heaven on Earth through positive action." The emphasis on UFOs as angelic emissaries may induce an eye-roll, but the internal logic aminating the One World Family's vision is the same as that of many utopians: People should eat in ways that align them with the forces of the universe.[14]

We now know that the macrobiotic diet isn't very healthful, but that doesn't undo the influence it had. In *Hippie Food: How Back-to-the-Landers, Longhairs, and Revolutionaries Changed the Way We Eat*, Jonathan Kauffman writes that "macrobiotics introduced some radical changes into the diet of the counterculture . . . [It] introduced legions of young cooks to Asian ingredients, helped bring stir-frying and steaming to the masses, and gave the first shape to an emerging cuisine built on the foundations of eating seasonally, organically, and often locally." Not only were young cooks developing a taste for Asian ingredients, they were also trying many novel spices, herbs, and combinations from a range of culinary traditions. Often purchasing their staples at ethnic markets, the hippies discovered new ingredients and flavors; in their recipes, Middle Eastern, Indian, and Mexican influences abound.[15]

The recipes Lucy Horton collected in her cross-country travel to various communes include dishes from many cultures—among them vegetable curry, couscous, ground nut stew, falafel, burritos, goat stew, huevos rancheros, dahl, poi, chutney, and something called "Torgerson's Mexican-Italian Blintzes." As that last suggests, mixing and matching across traditions was prevalent. One communard Horton met recalled that "before I lived communally, I could cook scrambled eggs and TV dinners. Now I cook curry, quiche, frittata, shark steak, crepes, and tamales . . . Living with people who enjoy eating anything and everything is a broadening experience."[16]

Eating with Crescent Dragonwagon would have been especially broadening. She devotes an entire section of her book *The*

Commune Cookbook to recipes for organ meats, not because they'd gained acceptance in American cuisine, but because they hadn't. She wrote "the vast majority (the silent one, you know) of us, while gladly sinking our teeth into a bloody rare roast or steak, pale at the thought of eating something's liver or heart or kidney or brain. This is our white middle-class heritage—we never grew up eating any of these things." Dragonwagon tried all the organs, describing her initial revulsion and the surprise of finding that she loved heart.[17]

With sections on making "stonies" and others on preparing wild animals ("old squirrels should be parboiled," for example, and "of course, you can eat snakes. Just cut off their heads, and skin them down the belly, then cook them like you would a porcupine"), its cultural breadth is far from the most striking attribute of *Eat, Fast, Feast: A Tribal Cookbook by the True Light Beavers*. Still, it's worth noting that it contains recipes for "'Jewish-Italian' bucatini, lamb curry, Puerto Rican *habichuelas*, spicy Lebanese coffee, and a sabayon made with Grand Marnier."[18]

Alice May Brock, who became famous thanks to Arlo Guthrie's 1967 song "Alice's Restaurant," archly assured readers in *Alice's Restaurant Cookbook* that they needn't be daunted by foreign fare. The chapter "Foreign Cookery" reads in its entirety: "Don't be intimidated by foreign cookery. Tomatoes and oregano make it Italian; wine and tarragon make it French. Sour cream makes it Russian; lemon and cinnamon make it Greek. Soy sauce makes it Chinese; garlic makes it good. Now you are an international Cook."[19]

What these cookbooks illustrate is of a piece with the messages in other cookbooks of the era: Ethical, political, and spiritual concerns were as important as economics and flavor in shaping diets. Vegetarianism enabled eaters to escape the bourgeois diets of their parents, industrial agriculture, and patriarchy; it was also kinder to the planet and to other mammals than was omnivory. That said, meat-eating could be an ethical option—if one carefully considered the well-being of the animal. And rejecting the mainstream diet offered an array of opportunities to broaden one's palate and to show solidarity across racial and ethnic divisions.

ALT.FOOD

UNLESS THEY HEWED TO a meat-heavy diet consisting of wild animals, these seventies-era back-to-the-landers needed to raise much of what they'd eat and needed to do so in a healthy way. Learning how was so empowering that one writer for the *Whole Earth Catalog*, Stewart Brand's influential compendia of tools, opined, "[I]f I were a dictator determined to control the national press, *Organic Gardening* would be the first publication I'd squash, because it is the most subversive. I believe," the author continued, "that organic gardeners are in the forefront of a serious effort to save the world by changing man's orientation to it, to move away from the collective, centrist, super-industrial state, toward a simpler realer one-to-one relationship with the earth itself."[20]

Lucy Horton saw this proposition put into practice during her travels. "Many people moved to the country because they wanted to eat organically and found that impossible to do in the city," she wrote. "As a result, they all garden extensively." Her observation resonates with the experiences of the seventies-era back-to-the-landers I know. Whatever their other reasons for turning to the land, all of them maintained a small farm or a sizable garden and saw it as essential to their ability to live simply and well.[21]

Levi Walton, who grew up in New Jersey, moved to Maine just a semester shy of graduating from college because he found the mainstream he was being channeled toward oppressive. When he moved north in the early seventies, it wasn't with the aim of growing his own food, but because he wanted to live lightly, have a fun life, and be free: "Suburban kids [were] realizing that the kind of lifestyle our parents were participating in was not something we were going to buy into because, well, it certainly lacked freedom." As did many other back-to-the-landers, he came to see growing his own food as a part of being free. He still recalls the exact moment he decided to garden: He was visiting a friend in the spring of 1974 and watched his buddy dig up parsnips he'd overwintered. Levi was amazed; he'd never seen anyone do that, nor had he ever tasted such sweet parsnips. "I've been gardening like crazy ever since," he said, first as a "squatting gardener" until he got a place of his own, then on his land. More than forty years later, he's still at it: "I'm still a firm believer in growing as many vegetables as I can."[22]

Jay Davis, who'd been working as a newspaper reporter in Boston, was inspired to go back to the land by civil rights issues; women and people of color were struggling to gain equality, and the system was clearly working against their well-being. Jay explained that he decamped for rural Maine because he wasn't "interested in a world where Blacks are being beaten and arrested, and where women aren't treated the way they should be. It just seemed like the values of America sucked and I just didn't want to be part of that." When he and his wife got to Maine, Jay was twenty-seven, which he said made him old compared to many other back-to-the-landers. At first, he had no idea how to live lightly: "I read the Nearings' book *Living the Good Life*, so I had some intellectual awareness of living off the land, but no hands-on experience whatsoever." Even so, he and his wife bought an old farm, heated it with wood, and "had a huge garden and raised animals, cut hay—kind of a homesteader lifestyle."[23]

Craig Wilgus didn't start out planning to have a huge garden or help to run a food co-op, both of which he ended up doing. Growing up, he'd never wondered about the foods he was eating. "But when you get out into the real world," which Craig did in the late 1960s, "it just seemed like there was a lot of information, and a lot of people asking questions. I'm sure the Vietnam War had a lot to do with that, because here was a war that people questioned *after* a war that people didn't. So if you're going to question one thing, why not question everything?"[24]

In 1969, while traveling in Hawaii, Craig became aware of "many influences floating around, including a lot of eastern religious stuff" and sensed that "rice kind of went hand-in-hand with that." He vividly recollected discovering brown rice there, meeting "this kid from California, and all he was eating was raw brown rice. He was always grinding away on it. He had to eat *all* the time." The kid inspired him. "If he can eat that stuff like that," Craig reasoned, "then I can certainly eat it cooked."

When Craig visited St. George, Maine, he felt as if he'd found the place he belonged; he bought a piece of land and built a tree house to live in. Then some friends found a bunch of steel pipe at the dump, and one of them figured out how to make a geodesic dome. Craig used the leftover pipes and made a dome, covered it in plastic, and lived in it for two years, while he built a more permanent house. He heated the dome with woodstoves, which worked out well because he was delivering wood at the time and his boss paid him in wood. Craig called it "easy and cheap living!"

He continued to question just how healthful his youthful diet had been, and resolved, as did so many other back-to-the-landers, to grow as much food as he could, to eat less meat and fewer processed foods, to avoid white sugar, and to incorporate more rice and more whole grains into his diet. His questions about the status quo weren't confined to food: He wondered about the wisdom of pursuing nuclear power, the rationale for not investing broadly in solar power, and, perhaps most of all, the benefits of a hurried, frenetic lifestyle. In its stead, he sought—and found—a "sense of community and family and the simple life."

Although environmental issues were the main reason Sherm Hoyt went back to the land, desires for a producer-oriented and community-oriented life were also important. On an autumn day in 2015, as we talked about his decision to live lightly on the land, Sherm traced his misgivings about capitalism to his family. His great-grandfather had owned several railroads and lived in a fifty-room mansion, but the next generations of his family didn't stake out their own paths. Instead, Sherm said with some distaste, "we just basically lived off my grandfather's wealth. We were really a family living off the work of others."[25]

Much like Scott Nearing and Frank Stephens, who viewed US involvement in World War I as motivated by corporate greed, Sherm saw the Vietnam War as influenced by economics: The "war machine was making a lot of people very rich," he observed, adding that "the economy was benefiting from the war." He was able to maintain a deferment and avoid serving in the Vietnam War by teaching, and in 1970 he organized the events at his school for the first Earth Day. That was a "critical point," as he was becoming more attuned to the environmental problems the world faced and more confident in his ability to teach others about them.

Around the same time, he began working at a small camp in St. George. Blueberry Cove Camp had been established in 1949 as one of the first interracial camps in the country, bringing together Black youths from urban areas with white youths from mostly rural ones. In 1973, Sherm moved to Blueberry Cove; he and some coworkers were reading *The Whole Earth Catalog* and searching for "a more sustainable way to live." For about a year, they pursued it together, but the group broke apart because, as he explained, "one person owned the land, and landownership is very complicated in group-living situations. There needs to be a way to share the ownership in a successful communal situation"—an observation many of his predecessors would agree with.

Moving out on his own, Sherm said he wanted to be a "dirt farmer." He still lives in the idyllic spot on the edge of a beautiful ocean cove where he moved then. Unfortunately, it's "a terrible place to grow food. The soil's very heavy and wet when it rains, and there's not enough of it." He amended the soil and planted vegetables; away from the water, he began grafting fruit trees and growing alfalfa hay, keeping goats and chickens. Despite his efforts, the land simply wasn't destined to be farmed, so Sherm turned his attention to the sea. He tried lobstering, clamming, oyster farming, and seaweed gathering. Lobstering could be a financially viable—some years even a lucrative—endeavor, but in the 1970s many of the other seafoods had a limited market. However, just as the hippies altered so much else about what we eat, they helped generate broader interest in seaweed farming and foraging. Long-line kelp farming is finally coming into its own in the United States and is contributing to the growth of the global seaweed market. Oyster farming, likewise, is on the rise.

When John Bly turned twenty-six and aged out of the lottery for the Vietnam War, he said he finally considered himself "free to start his own life." His first stop was in Massachusetts, where he cared for his ailing father. While he was there, a neighbor gave him a copy of the Nearings' *Living the Good Life*. Before he was even halfway through the book, he'd decided their good life was the life for him.[26]

In 1972, he came to St. George to fix the roof of his grandparents' house and ended up staying. Every weekend that summer, he went to Cape Rosier to help the Nearings and Eliot Coleman, splitting his time between their two farms. He helped the Nearings construct their stone house, mulch blueberries, whatever needed doing. (He became so enamored with working with stone that he eventually built a stone house for himself and his family.) During the workweek, he built himself a cabin and outhouse near his grandparents' old home. Like Craig's dome, the cabin had no running water or electricity and was lit by kerosene lamps. In 1974, John Bly met Chris, his wife to be. Both were passionate about living off the grid out of concerns about nuclear power. Chris said her "desire not to use electricity was to reduce the number of people who were using it so that nuclear would seem less necessary," adding, with a gently wry laugh, that she knew it was "absurd, in a way, that it wouldn't really make a difference. But I had to do it."

More generally, Chris wanted to take back control over her life; she believed people were being forced to rely too much on experts, often at their own peril. A nurse, she pointed to the pervasive use of anesthesia during childbirth and the proliferation of antibiotics as examples. Anesthesia made delivering a baby easier for the doctor but was often unnecessary, even undesirable, for the mother. And antibiotics were then so new that their drawbacks were unknown; even so, doctors often turned to them as the first option. She promoted home births and plant-based remedies as wiser choices—both of which were radical at the time. The Blys earned part of their income foraging seaweed and selling it to Maine Coast Sea Vegetables. They tried to earn the rest by farming organically, but in the seventies the market for organic goods was still too small in their area. All the same, they continued to grow food organically for themselves and their family.

Farther north, on the Blue Hill Peninsula, Eliot and Sue Coleman had better luck selling organic vegetables. They arranged them in baskets beside a chalkboard listing prices. Few people came at first, but "soon word got out," as their daughter Melissa recalled, "and the trickle of customers wandering down the grassy lane began to increase. They'd heard about the flavorful tomatoes that Papa pronounced 'to-mah-toes,' 'because you say *ahhhh* when you taste them.' The carrots were so sweet, they were eventually dubbed 'candy carrots' by an appreciative child."[27]

In their embrace of organic growing methods, these Maine-based back-to-the-landers were of a piece with their peers throughout the country. That shared emphasis was, as Lucy Horton wrote, "no coincidence . . . If you want to eat organically grown food—that is, food not chemically fertilized, sprayed or processed—you almost have to grow your own." Like their predecessors who feared food adulterants, these back-to-the-landers knew conventionally grown produce, grains, and meats could retain residue from chemical fertilizers and pesticides. The food safety laws of earlier decades were still on the books, but the number of chemicals used in agriculture had ballooned beyond the USDA's ability to test for them all.[28]

When Jeanie Darlington decided to grow a food garden, in 1968, she was working at a plant nursery a few towns north of Berkeley and "had plenty of knowledge about all the super fertilizers and magic bug killers." One "magic bug killer" label said it could be used on vegetables until one day before they were harvested. That claim got Darlington wondering: "If it kills all the bugs it says it does, how come one day will make it safe for me?" Knowing what she did about pesticides, she taught herself to garden "without poisons." In fact, as she learned more about organic gardening, she realized "a third of the products I was selling were only making Standard Oil richer and the air and earth more polluted." (She quit.)[29]

Among the most harmful pesticides in use then was dichlorodiphenyltrichloroethane—DDT. During World War II, DDT had been sprayed in tropical regions to kill malaria-carrying mosquitos; afterward, it was widely used as an agricultural pesticide. Then, in 1962, Rachel Carson documented how harmful DDT was, not only to insects, but also to frogs, fish, birds, and mammals. The impact of *Silent Spring* was so profound that it's credited with helping establish the modern environmental movement. Chemical companies were not pleased. They tried to quash the book; one even threatened to sue Carson's publisher if it was released. Others challenged her science or made personal attacks. But the industry outcry didn't deter her or those publicizing her findings. The year it came out, *Silent Spring* was serialized in the *New Yorker* and was a Book-of-the-Month-Club selection. The following April, Carson was a guest on a CBS *Special Reports* episode entitled "The Silent Spring of Rachel Carson," which was watched by more than ten million people.

Although Carson succumbed to breast cancer before seeing the full impact of her work, by 1969 "almost 60 percent of Americans surveyed said they thought that agricultural chemicals posed a danger to their health, even if used carefully." A meaningful minority began to reject foods grown using toxic pesticides, but the industrial model of agriculture remained—remains—firmly in place. And the chemical load continues to grow. More than a billion pounds of pesticide are used annually in the United States, twice as much as when *Silent Spring* came out.[30]

The consumer advocate Ralph Nader also warned people about food-related dangers. Rarely circumspect, he decried the behavior of food conglomerates, calling them "oligopolies" and accusing them of causing "erosion of the bodily processes, shortening of life, or sudden death." In his 1967 exposé "Watch That Hamburger," Nader described conditions in meat-processing plants in Delaware; it contains this passage so disgusting it calls to mind Sinclair's *The Jungle*: Vermin "had free access to stored meats and meat products ingredients . . . Dirty meats contaminated by animal hair, sawdust, flies, rodents and the filthy hands, tools and clothing of food handlers were finely ground and mixed with seasonings and preservatives. These mixtures are distributed as ground meat products, frankfurters, sausages, and bolognas." Elsewhere, Nader accuses food processors of disregarding the "nutritional, toxic, carcinogenic, or mutagenic impact" of their products "on humans or their progeny." And, much as Kallet and Schlink argued in *100,000,000 Guinea Pigs* more than thirty years earlier, he claimed the FDA knew industry giants pursued practices that endangered Americans and did little to stop them.[31]

BREAD MAKES A COMEBACK

As HAD SCOTT AND Helen Nearing, the seventies-era back-to-the-landers preferred to be producers rather than consumers, making not just their dinners, but also their homes, medicines, clothes, and pastimes from scratch. One quirky but significant way they differed from their iconic "grandparents" was that many of them liked bread. So much so that for the first time since WWII rationing, bread-making enjoyed a serious revival. The kids raised on Wonder Bread wanted to learn to make alternatives, and they invested those loaves with social and political meanings as well as flavor. "I used to make a lot of bread," remembered Maine-based back-to-lander Craig Wilgus. "We all had those little Corona mills, and we were out there grinding away . . . I only eat whole grain breads—the grainier the better. That has stuck with me. I can't eat white. White flour, white sugar."

Many of his peers also made that brown-versus-white distinction; in fact, the food scholar Warren Belasco called it the "central contrast" between hip and mainstream diets. Whole grains are more nutritious than bleached white flour, but brown foods were so pervasive in the hip diet in large part because they served as a metaphor for political solidarity with people of color. Malcolm X maintained that "if you want bread with no nutritional value, you ask for white bread. All the good that was in it has been bleached out of it, and it will constipate you. If you want pure flour, you ask for dark flour, whole-wheat flour. If you want pure sugar, you want dark sugar." Similarly, a writer for the underground newspaper *Quicksilver Times* denounced white flour, sugar, and rice as "bleached to match the bleached-out mentality of white supremacy." White food became both a metaphor for the system that produced it, a nefarious "regime of experts and technocrats who, for the sake of efficiency and order, threatened to rob us of all effort, thought, and independence," and a metonym of it, a "dangerous drug" contributing to our collective downfall, leaving legions of eaters weak, lacking in vitality, and "faded" in appearance.[32]

Communard and cookbook author Crescent Dragonwagon recounted how selling wholesome bread furthered her understanding of capitalism. She and a housemate enjoyed making bread and were good at it. So when someone asked to buy sixteen loaves a week, they said yes. But "we figured out the cost of the breads, and it was . . . really high compared to store prices, even for healthy bread, and this not even counting 'profit.' It was impossible to do it. It worked out that I was getting something like two dollars for a whole day's work." For Dragonwagon, the experience confirmed her sense that "*capitalism can't work.*" At the same time, it led her to a deeper understanding of the political potential of food, as she realized that "baking a loaf of brown bread in this society is revolutionary, if you know why you're doing it."[33]

Most back-to-the-landers stuck to making more manageable quantities, baking mainly for their households. Melissa Coleman, one of Eliot Coleman's daughters, recalls sitting on a wooden stump that served as a kitchen stool, enjoying the smell of bread baking in the family's wood-stoked cookstove. Her mother, Sue, let Melissa help make chapati: "Mama let me help mix the flour from the grain mill with water and salt to make a pliable dough," she remembered, "then Mama kneaded it to bring out the gluten and let it set for an hour before making round golf balls of dough that she flattened with a rolling pin into thin, but not too thin, pancakes." After searing them on both sides, her mother placed each round "on a bent clothes hanger over hot coals inside the firebox, where it would blow up into a steamy balloon. Once it was removed from the flame, the air in the middle was released and the balloon flattened to form a perfect tortilla-like vehicle." Chapati wasn't common in white Americans' kitchens then; Sue learned to make it from David

Hatch, co-owner of the health food store in Vermont where she and Eliot first encountered the Nearings' *Living the Good Life.* He'd learned to make it in India.[34]

Back-to-the-landers could find recipes and bread-making instructions in books that ranged from 1962's groundbreaking *Let's Cook It Right!* by Adelle Davis to Lappé's *Diet for a Small Planet* and Mildred Ellen Orton's *Cooking with Wholegrains*, both from 1971. More bread recipes could be found in the *Whole Earth Cook Book: Access to Natural Cooking*, by Sharon Cadawallander and Judith Ohr, in Sandra and Bruce Sandler's *Home Bakebook of Natural Breads & Goodies*, in Horton's *Country Commune Cooking*, and Dragonwagon's *The Commune Cookbook*, all of which came out in 1972. Two years later, aspirants might have turned to *Cosmic Cookery.* Some folks probably relied on the recipe for the Diggers' coffee can bread, which appeared in the first issue of *The Mother Earth News*, or on one of the recipes in *The Whole Earth Catalog.* Arguably, *the* bread-making book of the era is Edward Espé Brown's Zen-infused *The Tassajara Bread Book*, which came out in 1970. *Tassajara* became a staple, selling three-quarters of a million copies. Brown incorporated plenty of gentle, nurturing reassurances to guide nervous bakers as they struggled to make non-bricklike loaves.

The counterculture's emphasis on eating grainy, whole breads gradually permeated the mainstream. Sales of white breads fell and those of whole grain breads rose; however, like the faux foods the FDA hadn't been able to regulate in the 1930s, the "wheat breads" many companies sold were really white bread with a dollop of whole wheat or wood pulp added in. (Bred-Spred, with its hayseeds simulating berry seeds, sounds like the perfect "jam" to put on a slice of wood pulp "whole wheat" toast.)

Before long, hippies and former hippies were establishing bakeries featuring traditional breads, like the People's Company Bakery, in Minneapolis; the Women's Community Bakery, in Washington, D.C.; Slice of Life Bakery, in Cambridge, Massachusetts;

Dharma Crumbs Bakery, in Colville, Washington; and the Yeast-West Bakery, which was in Buffalo. Many novice bakers traveled to Europe to learn from traditional bakers; "returning home, they showed Americans what real peasant bread tasted like." The rise of artisanal bakeries led Tom McMahon to establish the Bread Bakers Guild of America, in 1993, to bring the many makers of rustic loaves into community.[35]

Store-bought bread continued to dominate in the mainstream, but people who hadn't been hippies also caught the bread-baking bug. In the late eighties, almost two decades after *The Tassajara Bread Book* first hit the shelves, I bought a copy of the then recently revised, updated version for my husband, who was eager to learn to make bread. The stains in Rob's copy suggest that in the 1980s, he made only yeast breads. Later, he replicated the journey many homesteaders made, switching from using yeast to maintaining his own sourdough starter. The seventies-era homesteaders' and communards' attention to breads and bread-making helped usher in the artisanal bread craze of the 1980s and nineties, which has morphed from a craze to something closer to a staple; on its website, the Bread Bakers Guild lists close to five hundred artisanal bakeries across the United States.

During the first spring of the COVID-19 pandemic, as folks began making sourdough bread as a source of solace, I felt extra grateful for the mostly unsung contributions of the back-to-the-landers and hip bakers who reintroduced grainy flours and sourdough starters to the American palate.

YEARNINGS FOR CHEESE

WHEN LUCY HORTON WAS crisscrossing the United States, hitchhiking to each next commune, she managed to find ways to bring her hosts small thank-you presents: "Usually I would bring some contribution . . . I would buy cheese somewhere, then

I would bring it as a gift." I suspect her gifts were widely appreciated. By the time the hippies went back to the land, "real" cheese had pretty much disappeared from American stores.[36]

In the United States, artisanal cheesemaking started to wane late in the nineteenth century, when factory cheese was introduced. During World War II, the few remaining regional cheesemakers were acquired by industrial makers as part of the homeland contribution to the war: "streamlining commodity cheeses in factories was an important wartime effort and a way of saving money while providing cheese for the nation. The cheese business was consolidated, for the sake of winning the war." After that, most Americans dwelt in what Murray's Cheese in Greenwich Village described as "a desolate wasteland of 'cheese food' and 'cheese product.'" Back-to-the-landers weren't interested in "cheese product." Nor were they close enough, or financially flush enough to afford, to buy imported cheese from places like Murray's Cheese and DeLuca's Cheese Store, both in Greenwich Village—though hip folks close to Berkeley might have gone to the Cheese Board, a collectively run cheese shop that's still thriving.[37]

Generally, back-to-the-landers who wanted real cheese could assuage their yearnings in one of two ways. Quite a few made it themselves. Although that may sound daunting, the other way, which we'll look at in the next chapter, was considerably more complex: They created an alternative food system through which they could buy and sell real cheese (and many other things).

Making great cheese is an exacting art but making an okay edible cheese isn't too difficult, especially if the cheese doesn't require aging. However, cheese-making usually requires raw milk, the sale of which was—and still is—illegal in many states. Self-sufficient homesteaders solved that problem by keeping their own milk sources, occasionally cows but more often goats. In *The United States of Arugula*, David Kamp dubbed goats "the ultimate hippie livestock, both because they evoked the humble peasant cultures that the counterculture romanticized and because they were easy to care for." So easy, in fact, that 28 percent of the more than five hundred back-to-the-landers Jeffrey Jacobs surveyed while working on *New Pioneers: The Back-to-the-Land Movement and the Search for a Sustainable Future* reported having raised goats.[38]

Goat cheese wasn't popular in the United States "[o]utside of countercultural and Francophile gourmet circles," as Kamp wrote. Nor was goat yogurt. Redwood Hill Farm, in Sonoma County, California, claims its goat yogurt was the first commercially available in the nation—in 1982. Well before that, hippie communards and homesteaders interested in making goat-milk cheeses and yogurt were finding guidance in *Mother Earth News*, the go-to periodical for the movement. Its second issue contained an article on raising dairy cows and goats. Over the remainder of the decade, *Mother Earth* published quite a bit more about goats, including articles on making goat cheese and yogurt, on Andalusian-style goat cheese, on hard goat cheeses, on goat cheese ricotta, and on goat milk butter.

Such reading material was enough to get home cheese and yogurt makers up and running, but not enough to help those who'd become passionate cheesemakers take their efforts to the next level. And quite a few back-to-the-landers wanted to do exactly that. Among the many who set out to get great at making goat cheese were Laura Chenel, Mary Keehn, Caitlin Hunter, and Bob and Ricki Carroll. Their successes introduced Americans to goat cheese and helped launch an artisanal cheese scene that's still flourishing. Americans now eat three times as much cheese and they did in 1970 (albeit not all of it artisanal).

Having gone back to the land in California during the early 1970s, Laura Chenel bought a goat. Soon, she discovered she was crazy about the creatures: She said it was like "I belong to the goats, that I am in their tribe." As her herd grew, so, too, did the amount of milk it produced. Chenel "felt responsible for the milk and wanted to transform it into something that would pay tribute to its makers; therefore, she found herself drawn into cheesemaking." Chenel learned to make goat cheese, but when she looked for a mentor to help her become a professional cheesemaker, she couldn't find anyone in the United States. So, in 1979, Chenel traveled to France, where she spent time with four different farmstead families and learned to make great cheese. Fortune favored Chenel; among her first clients was Alice Waters, creator of Chez Panisse: "Waters fell in love with Laura's cheeses, and decided to place a standing order for the restaurant (50 pounds a week). At Chez Panisse, Alice breaded and baked slices of Laura Chenel chèvre (French name for 'goat') and nestled them in beds of mesclun greens, creating the iconic 'California Goat Cheese Salad.'"[39]

Mary Keehn, a self-described "serious hippie" who didn't think much of the "slices of orange, rubbery, plastic squares" that passed for cheese, lived in Humboldt County, a few hours north of Chenel. She began her herd with two goats from a neighbor. As the herd grew, she tried to perfect a goat cheese to feed her daughters. Later, like Chenel, Keehn traveled to France to learn more about making cheese so she could turn her home-dairying effort into a business. On the plane ride back from Europe, "a prophetic vision danced across Mary's imagination. She awoke with a clear image of a new kind of cheese: a cheese accented with a thick line of ash reminiscent of the fog often blanketing the expansive Humboldt County coastline—and the idea of Humboldt Fog was born." She launched Cypress Grove Chevre in 1983. The artisanal cheese company features not only Humboldt Fog, but also other cheeses whose hip origins are reflected in their names, including Purple Haze, Sgt. Pepper, and PsycheDillic.[40]

By 1983, the sorry state of American cheese production had improved enough to prompt Frank Kosikowski, a dairy science professor and international cheese expert, to establish the American Cheese Society, "as a national grassroots organization for cheese appreciation and for home and farm cheesemaking." In

1985, at its annual conference, the group instituted a cheese competition. That year thirty makers submitted seventy cheeses; in 2019, 257 makers entered 1,742 entries! Now ACS has more than two thousand members, almost half of whom are professional cheesemakers. Opportunities to buy real cheeses are no longer limited to those who live within the orbit of high-end cheese shops. Thanks in great part to those hippie-era cheesemakers, delicious artisanal cheeses are now a mainstay of farmers' markets, food co-ops, and even some grocery chains.[41]

Much as those back-to-the-landers were interested in making it themselves, foodies and fans of self-sufficiency also seek out opportunities to learn to make cheese. For years, I tried to teach myself. My aspirations weren't as lofty as Chenel's or Keehn's; I just wanted to make a decent chèvre and maybe a hard cheese like cheddar. But simply following written directions yielded sad results; I needed to be taught.

I took my first workshop in April 2008 with Ricki Carroll at the New England Cheesemaking Supply Company, which is in western Massachusetts. At that time, she'd recently become well known outside the cheese world because Barbara Kingsolver described one of her classes in *Animal, Vegetable, Miracle*. All but three of the folks in my workshop had learned about Ricki from Kingsolver's book. Laughing, Ricki told us her company straddled two eras, BB and AB—*Before Barbara* and *After Barbara*.

Long BB, indeed, long before Ricki became acclaimed as "the Cheese Queen," she and her husband were hip communards raising vegetables and chickens and a couple of goats in Ashfield, Massachusetts. As had happened to Laura Chenel and Mary Keehn, the Carrolls' goats produced more milk than they could drink. So, knowing absolutely nothing about how to do it, they decided they should learn to make cheese. Their search led them to dusty corners of the library, where they found books "of 1800's vintage" that had recipes and explained the process. Finding mentors and supplies proved more difficult.[42]

In 1978, the couple "wrote to the embassies of just about every country in the world that [they] thought might have an industry producing small-scale cheese-making equipment," and eventually found a farm family in the south of England who made wonderful cheeses and handcrafted cheese presses. Before departing for England, the Carrolls took out an ad in *Dairy Goat Journal* offering catalogs of cheesemaking supplies for twenty-five cents. When they got home, they were pleasantly shocked "to find their mailbox overflowing with quarters." Within a few months, Ricki and Bob were selling supplies and teaching workshops in their home kitchen.[43]

By the time Ricki taught the introductory workshop I attended, she'd traded students standing stove-side in her home kitchen for a mirrored, stainless-steel demonstration kitchen that looked a lot like a science lecture hall. Even so, the workshop retained a groovy vibe and an infectious delight in all things cheese. In just six hours, she showed us how to make a basic hard cheese, yogurt, crème fraiche, mascarpone, queso blanco, whole milk ricotta, whey ricotta, and her signature thirty-minute mozzarella. For years, Ricki's basic hard cheese was my go-to recipe for making aged cheese.

Though Ricki started out making goat's-milk cheeses, she used only cow's milk in our workshop. I still really wanted to learn to make chèvre, and in 2009, serendipitously, Caitlin Hunter, who operated Appleton Creamery in Appleton, Maine, until 2021, began to conduct workshops. Her cheeses were very popular, so spring through late fall for the creamery team was a whirl of making and selling, much of the latter at regional farmers' markets. But by mid-winter, when the goats were dry, when the farmers'-market schedule slowed, Caitlin Hunter had some time to teach amateurs about her craft.

As she introduced us to the goats, I could see why our mid-March workshop was the last one that year: The ladies were tautly pregnant, closing in on their kidding dates. After we met them,

Caitlin herded the workshop's six students, all women, into an impeccably clean commercial kitchen lined with stainless-steel sinks and counters, a humongous stove, and dozens and dozens of cheese molds, cheese cloths, utensils, herbs, and spices. We made chèvre, feta, mozzarella, and ricotta. Caitlin even shared the secret for one of her most popular cheeses, a chèvre packed in olive oil with roasted garlic and herbs—so good it's won American Cheese Society awards three times. When I make chèvre, I still use her instructions.

Cheese, whole grain breads, rice, ethnic cuisines, organic produce, less meat-centric and sugar-centric diets. The seventies-era back-to-the-landers broadened the American diet, not as their initial aim, but as a by-product of their quest for freedom and a more authentic lifestyle. As their gardens and recipes demonstrate, they saw food and ethics as intertwined. They wanted to eat in ways that nurtured the body and spirit and planet, to enjoy foods that were delicious and consistent with their commitment to equality. In homesteads and communes, this generation relearned lost cooking skills and reintroduced their fare to the wider world. Even more remarkable, they created an entire food system to accomplish their goals, one that continues to operate as an alternative to America's industrial agricultural system.

8

TRANSFORMING THE FOOD SYSTEM
FROM SEED TO TABLE

CONTEMPORARY YOUNG FARMERS, THE fifth wave of American utopian back-to-the-landers, are nurturing their communities through their agrarian work, creating integrated regional agricultural economies, and generally making locally grown, often organic, non-GMO foods more widely available. In its most recent five-year strategic plan, the National Young Farmers Coalition is steely-eyed about the urgency of the situation, writing, "We believe we have one generation to transform agriculture in service to our communities and the land. We are that generation." Their vision is of "a just future where farming is free of racial violence, accessible to communities, oriented toward environmental well-being, and concerned with health over profit." Their tasks, both on farm and off, to try to replace the broken industrial agricultural system, are herculean. Americans need them to succeed; among the many good reasons for pursuing these changes, we need to dramatically reduce agriculture's contributions to climate disruption.[1]

Daunting as their situation is, at least they aren't starting from scratch. Thanks to a host of seventies-era hippies—many who went back to the land as well as some who didn't—the current generation entered an alternative agricultural world with the

essential infrastructure already in place. Building those founda-
tional pieces wasn't easy. Nor was it a unified process; rather, it
was an amalgam of parts and efforts and interests that slowly
accumulated into a coherent alternative system for growing and
distributing organic foods and other goods. Unlike the dominant
model, the alternative was fueled not by corporate capital, but by
sharing—of information, resources, tools, and seeds. And again
unlike the dominant model, it had few supporters in high places.

Nationally, sentiment against organic agriculture during the
1970s was powerful—and reinforced by people in power. No longer were granges and the Farm Bureau the most persuasive agricultural lobbies. Instead, commodity growers and processors held
sway. By the end of World War II, the politicians who represented farmers were representing their own interests almost exclu-
sively. The House Committee on Agriculture, for example, "was
dominated by ten commodity subcommittees that drew membership from congressional representatives whose districts grew
mainly those particular crops." Legislators vied for benefits for
their home districts, and their colleagues understood they were
expected to go along with the requests. The power of legislators
from regions growing commodity crops became more entrenched
over the successive decades. By the time back-to-the-landers were
growing organically, it "bore a highly negative connotation in political circles."[2]

In 1971, Earl Butz (the Secretary of Agriculture who told farmers to "get bigger, get better, or get out") sardonically dismissed
organics, saying: "Before we go back to an organic agriculture in
this country, somebody must decide which 50 million Americans
we are going to let starve or go hungry." Under Butz's leadership,
the USDA advocated chemical-intensive, monocultural farming.
Agriculture departments in universities and their Cooperative
Extension services followed suit. In the United States, few books
about organic gardening—and even fewer about organic farming—were available, though that would soon change.

To learn about organic methods, Eliot Coleman borrowed old books from his neighbor Scott Nearing. Coleman valued the lessons he gleaned from them so much that he calls them his "grandparents," a stand-in because he "didn't have grandparents in farming" who could educate him. Along with old books, information about organic methods could be found in a few contemporary magazines, such as Rodale's *Organic Farming and Gardening*, *Mother Earth News*, and the *Whole Earth Catalog*, the huge—and hugely significant—publication Stewart Brand started publishing in 1968. Roughly eleven by fourteen inches, printed on pulpy newsprint, a *Whole Earth Catalog* clocks in at more than an inch thick and two and a half pounds. Those oversized pages were packed with information—from instructions for tree pruning, to geodesic dome designs, to home cures for all sorts of ailments, to Hopi creation stories, to recipes for yogurt, to suggestions on canoe design. The content was meant to help "the individual to conduct his own education, find his own inspiration, shape his own environment, and share his adventure with whoever is interested." Essays, instructions, reviews, cartoons, terrible jokes, and excellent lists of resources are so densely presented that each bit must vie for the reader's attention.[3]

Almost every issue featured a cover image of the earth as seen from space. My copy, from June 1971, presents "probably the first American photograph of the 'whole Earth'"—the crescent of planet evident, the rest in shadows, was taken from Apollo 4 on November 9, 1967. Brand envisioned the catalogs as a new information system for a new world, one that recognized itself as a whole and profoundly integrated system. He hoped *Whole Earth* would help disparate communes and homesteads connect with one another, and his hope was realized. More than two million catalogs were sold in the first few years of its publication. It helped readers clarify and reinforce an engaged environmental ethos, a "small is beautiful" philosophy, and a reveling in self-sufficiency. That focus on personal power was encapsulated in the catalog's statement of purpose: "We are as gods and might as well get good at it."[4]

As valuable as the *Whole Earth Catalog* was, it couldn't fully replace the benefits that come with in-person exchanges. Across the nation, back-to-the-landers serious about growing food organically began coalescing into organizations to teach and learn from one another and to educate the public. In 1969, Samuel Kaymen, a thirty-five-year-old back-to-the-lander from Brooklyn, was offered a plot to farm in New Hampshire. Knowing absolutely nothing about growing food, he borrowed all the books on gardening and farming the local library had (and discovered none of them had been checked out in almost two decades). Realizing that he and the other novice farmers needed "a way to bring people together to share information, to learn from each other," Kaymen papered parts of Vermont and New Hampshire with notices about an organization he was forming, the Natural Organic Farming Association, inviting people to meet on June 7, 1971. A few older farmers came, along with more than a hundred young homesteaders and communards. Attendees left with a plan to sponsor seminars and seed exchanges, arrange for bulk shopping, create an apprenticeship program, and issue a quarterly newsletter. NOFA grew quickly and became an organic wholesaler of sorts, supplying food to eighteen co-ops in New York City.[5]

Although this was a great way to bring whole foods to city folk, some NOFA members feared it veered too far from their initial mission. In 1974, Robert Houriet, by then a strong voice in the organization, proposed a dramatic reorientation for the group: He urged getting out of wholesaling and creating a network of farmers' markets across Vermont to enhance the viability of small-scale farming in the Green Mountain State.

Out on the West Coast, CCOF, the California Certified Organic Farmers, was founded in 1973. The following summer, organic farmers and growers attended a symposium titled "Agriculture for a Small Planet," in Spokane, Washington. There,

Wendell Berry warned the audience about the consequences of a continued decline in the farming population: "If we allow another generation to pass without doing what is necessary to enhance and embolden the possibility of strong agricultural communities," he told attendees, "we will lose [them] altogether. And then we will not only invoke calamity, we will deserve it." Days later, Berry sent a letter to some of the people he'd met at the conference, asking, "From where you are, can you see any possibility of another kind of agricultural symposium . . . one that would try to bring together the various branches of agricultural dissidence and heresy?" They could. And did. And Tilth was born. Tilth's core group worked quickly and well, organizing a conference for that November, which was attended by more than eight hundred agriculturalists from all over the country.[6]

In the late sixties and early seventies, back-to-the-landers who moved to Maine often contacted the University of Maine's Cooperative Extension Service, which offers courses and runs workshops to provide agricultural skills and knowledge to growers, farmers, youths, and others. But for hippie growers, finding the information they needed through Extension was difficult. In 2015, Les Hyde, who had worked as an Extension Agent beginning in the mid-seventies, explained to me that most of his colleagues in that decade were anti organics, that Extension was complicit in "creating this mechanized food system we had." Drolly, he explained his aim had been to transform the agency from within: "I was determined that in my role at Extension, I was going to be part of the solution—and more mechanized was not the way to go." Though decidedly in the minority, Les wasn't the only Extension employee to see farming from a different vantage. He mentioned a colleague who'd become a key ally to the hippie growers. That unlikely prime mover was Charlie Gould.[7]

Gould entered the agricultural world poised to be a full-throated advocate of the new industrial methods. He earned his master's degree in agronomy from Rutgers University shortly after

World War II and began his professional life working for E. I. du Pont. Then, in 1965, Gould shifted careers; he became a Cooperative Extension Service educator for the University of Maine, a position much like the one Les Hyde landed a decade later. At first, Gould received just a few calls a year from people with questions about organic farming; like other Extension agents, he knew nothing about that method. By 1971, he was getting so many inquiries that he organized a meeting to get the information seekers together in one place. Wisely, he invited Scott and Helen Nearing. Between fifty and sixty people showed up, and that December, many of the attendees formed the Maine Organic Food Association, soon changing the name to the Maine Organic Farmers and Growers Association (MOFGA). Charlie Gould didn't feel the general disdain most Extension agents had toward organics, or toward back-to-the-landers. In fact, much as Sir Albert was converted to organic methods by the farmers he'd been sent to India to retrain, Gould became a passionate organics advocate who later insisted he learned far more from the hippies than they did from him.[8]

Despite the widespread enthusiasm among early organics growers, the retail market for organic foods remained minuscule. If they were going to succeed, they needed a bigger share of Americans' food dollars—and to get it, they had to establish a market for their products and mechanisms for getting their wares into their customers' hands. As if heeding the Diggers' admonition that folks create the world they want to live in, members of NOFA and of MOFGA, of Tilth and of CCOF, began building on the information they found in new journals and old tomes. Sometimes working in concert, sometimes solving the same problems independently, they gradually established an alternative to industrial agriculture and massive grocery stores, one the current generation of young farmers continues to expand and refine.

SOIL, SEEDS, AND TOOLS

MUCH LIKE THE CHEESEMAKERS Laura Chenel and Mary Keehn, who became interested in selling cheese only after years of making it on a small scale for home use, few back-to-the-landers initially intended to farm commercially. The vast majority were seeking self-sufficiency for their household or commune, growing food organically because that was the only way to ensure it really was. Moreover, so few places had an adequate market for organic produce then that the back-to-the-landers who did want to sell produce often struggled to do so. The upside to that lag in demand was that these novice growers had time to secure all the inputs for growing produce organically (notably healthy soil, organic seeds, and appropriate tools), and to become competent producers before they marketed their wares.

Like their predecessors, the seventies-era experimenters had to infuse the soil with more fertility to make it healthy—or suffer the consequences, as often happened. Plenty of hippies had experiences that more closely paralleled those of the Fruitlanders than those of the Ardenites. Even those lucky enough to start out with actual farmland, like Jay Davis, needed to amend the soil, to add back the nutrients years or decades of previous harvests had drawn from it. Others faced the more daunting task of starting on raw land. Eliot Coleman, for example, started out with a wooded lot. He had to chop down the spruces and firs, stump the newly formed field, and remove the many, many rocks that grow in New England fields. Every year, he added seaweed, horse manure, and hay as amendments.

The seventies-era back-to-the-landers could amend the soil using the same methods as did earlier generations, but they had a far harder time acquiring organic seed than did their predecessors. For most of America's history, growers had easy access to seeds. As far back as the colonial period, farmers and gardeners could buy from small, regional seed companies. And the nation's founders were such enthusiastic agriculturalists that they integrated their passion into government programs. Beginning in the early 1800s, "The Secretary of the Treasury initiated a program requesting U.S. ambassadors and military officers to gather seeds and seed data from their posts around the globe." And in 1839, the US Patent Office created a division tasked with collecting and distributing free seeds to farmers.[9]

In 1862, when the USDA was established, it took over the Patent Office's popular seed program, which was then distributing more than two million packages of garden and flower seeds each year. Ready access to free, high-quality seeds enabled farmers to "conduct extensive seed breeding and provide the genetic foundations for American agriculture. Farmers developed steady genetic improvement mainly . . . [by keeping and then planting] seeds from the healthiest and most productive plants." Using this traditional breeding method, farmers developed many variations of important food crops, each suited to its own micro-climate: some were heat resistant, for example, or drought tolerant, or faster growing. By the late 1800s, the much-loved program accounted for a third of the USDA's budget.[10]

Seed sellers were not fans; they considered the government giveaway an impediment to their own flourishing. They formed the American Seed Trade Association, whose main goal in its early years was to end the USDA's seed program. In 1923, they got their way. A few years after that program's demise, Congress passed the Plant Patent Act, which allowed individuals or companies to patent plants that reproduced non-sexually but prohibited them from patenting plants that reproduced via seed. The logic undergirding the distinction was that no one should be able to hold a patent on a life-form; plus, they worried that if patents were allowed, the introduction of new seed types might be curtailed.[11]

In 1970, Congress passed the Plant Variety Protection Act (PVPA), pretty much reversing course. Although the PVPA doesn't permit patents, it allows individuals and companies to obtain "Certificates of Protection" for "novel, sexually reproducing plant varieties." These certificates give the maker sole rights for twenty to twenty-five years (much as pharmaceutical companies have sole rights to a drug for twenty years under patent law). In essence, the certificates are patents-lite. Once sole rights to a seed could be had, even temporarily, spending on research and development made financial sense at industrial scale.[12]

Large corporations rapidly bought seed companies; within a decade, "most of the small seed firms vanished, as mergers and acquisitions created a new seed industry structure dominated by large companies with private investments." Gone were regional seed sources. Gone were those selling quirky varieties on a small scale. Gone, in fact, were seeds for 93 *percent* of the agricultural genetic diversity that was present in the United States in 1900. Making matters worse, the consolidated companies (for the most part) weren't interested in selling seed as such. Rather, the seed companies were being acquired by chemical and pharmaceutical firms interested in turning seeds into permanently patentable commodities.[13]

Fortunately, Rob Johnston Jr. felt the tug of soil and seeds. A slim, ginger-mustached member of the University of Massachusetts's freshman class in 1969, Rob helped start a food cooperative and a natural foods store in Amherst. After his sophomore year, he left college to work at Erewhon Farm, in New Hampshire. Most of the produce from the fifteen-acre farm went to Erewhon, the macrobiotic food store in Boston that Michio and Aveline Kushi had founded. Johnston recalled being asked to grow some Asian specialty vegetables and having to get the seeds from Japan. He became fascinated by seeds, and as his acumen grew, his reputation with growers did as well; people began turning to him when they couldn't locate seed for particular varieties.[14]

In 1973, Johnston founded Johnny Apple Seeds. The seeds he bred and sold were for short-season crops that could be grown organically, among them heirloom varieties he wanted to keep available. Initially, he worked out of his parents' house, in Acton, Massachusetts, a stone's throw from Thoreau's old stomping grounds and a short drive from Fruitlands. In 1974, Johnston moved to Maine. His reputation spread quickly; in the inaugural issue of *The Maine Organic Farmer & Gardener*, MOFGA's newspaper, the editor hailed Johnny Apple Seeds as "the first new age seed company." Johnston changed the company's name to Johnny's Selected Seeds, as someone else had trademarked Johnny Apple Seeds. As Johnny's Selected Seeds, the company continues to thrive, providing seeds to growers around the world—and beyond: In 2015, astronauts on the International Space Station grew Johnny's "Outredgeous" romaine lettuce.[15]

Frustrated by the dearth of clean seed, others also pursued routes like Rob Johnston's. On the other side of the country, Steve Solomon traded his typesetting business in Los Angeles for a homestead in Oregon. A self-described "capital-O Organic gardener with capital-O Opinions," Solomon established seed-trial plots in the Pacific Northwest's clay-rich soil, persuading friends to maintain some in their yards so he could keep the cultivars isolated. In 1979, he began the mail-order seed company Territorial. The company grew so quickly that Solomon, who'd begun it with a yearning for simple living, sold it in 1985. (The couple who bought Territorial have maintained it with a spirit akin to Solomon's.)[16]

These commercial efforts were complemented by a renewed attention to seed-saving. Diane Ott Whealy and Kent Whealy didn't see themselves as hippies, but they had some of their values: "We did believe in the virtue of living modestly off the land, and we did question the controlling social, economic, and political principles of our society, so in a sense we were antiestablishment" as Diane put it. When the couple and their two children began homesteading, in Missouri, they planted seeds for deep purple morning glories and pink beefsteak tomatoes that were descendants of seeds Diane's great-grandparents brought to America in 1884. Her grandpa Ott had routinely given away seeds, so Diane was startled to learn she was the only one who still had some when he died.[17]

In her memoir, she wrote that around 1974, she and Kent began seeing a lot of articles about declining genetic diversity, particularly for food crops. Kent wrote letters to *Countryside*, *Mother Earth News*, and *Landward Ho* looking for other people interested in seed-saving. They heard back from a few folks, collated the seed information they received, and sent out copies of the *True Seed Exchange* to all twenty-nine members of the group. Among the items in that first "catalog" were Paul Bunyan beans and Bird Egg beans, both sent in by a woman named Lina Cisco, who said both beans had been grown by her family since the 1880s. With pink, beige, or scarlet skin flecked with scarlet dots and dashes, Lina Cisco's Bird Egg beans are densely creamy when cooked; they're my current favorite for Cajun beans and rice.[18]

The seed-saving group grew quickly. By the time the 1976–77 *True Seed Exchange* reached the mailboxes of its three hundred members, the publication had expanded to include a few commercial resources for seeds as well—citing Johnny's Selected Seeds as "a source of good seed." In 1979, the group changed its name to Seed Savers Exchange, as it's still known, and began an actual set of seed banks in addition to acting as a clearinghouse for person-to-person seed exchanges. The Seed Savers Exchange saw itself as a "fail-safe"—a store of genetic biodiversity that could be tapped if something dire happened to the USDA germ facilities because, as Diane Whealy noted, government repositories were so underfunded they struggled to fill their mission.[19]

Seed Savers Exchange is now the largest nongovernmental seed bank in the United States, holding more than twenty thousand varieties of at-risk seeds. But it isn't the only one. Soon after the Whealys began the True Seed Exchange in the Midwest, another couple began a seed-swap group in Virginia's Tidewater region. Eager to be self-sufficient and frustrated by Vermont's short growing season, biologist Jeff McCormack and his wife, Patty Wallens, had moved to Charlottesville in 1977. Jeff McCormack soon became involved with a local seed-saving organization, learning from its members about quirky varieties that were no longer available, like potato onions. As the couple learned about other unusual varieties, they created the Southern Exposure Seed Exchange (SESE). Their first catalog contained just sixty-five offerings, including "Tappy's Finest tomato, Yellow Potato Onions, and Calico Crowder southern peas." By the next year, they doubled their offering.[20]

Among SESE's best-known items is Radiator Charlie's Mortgage Lifter Tomato, one of many tomatoes named "mortgage lifter" developed during the Depression. A resident of Logan, West Virginia, M. C. ("Charlie") Byles had a radiator-repair shop and kept a small kitchen garden. He decided to try to cross a German Johnson Brandywine with other tomato types, keeping the tomatoes he liked best for subsequent crosses. After six or seven years, he had a stable variety he loved, and sold its seedlings for a dollar each. The tomatoes were so popular that Byles paid off his six-thousand-dollar home mortgage in just six years. I'm not surprised by the high regard R. C. Mortgage Lifter earned. It's a great tomato; the fruits are heavy and lobed, often two pounds apiece. The meaty flesh is a deep, earthy red as dark as the skin, and the flavor is simultaneously sweet and spicy, with a suggestion of smokiness. The top tends to crack, making it less than ideal as a commercial variety, but for home growing it's stupendous. Even so, R. C. Mortgage Lifter is now listed on Slow Food's Ark of Taste, the international compendium of unique foods at risk of becoming extinct.

These four organizations were the vanguard of what became a modestly sized, but very real, movement of heirloom and organic seed sellers and seed-savers determined to preserve biodiversity and meet the needs of homesteaders, organic farmers, and organic gardeners. Thanks to groups like these, contemporary young farmers have ready access to a wide variety of organic seeds.

But even these good efforts can't ensure that all the seed persists. Will Bonsall, the seed saver who gave the Maine Grain Alliance its first pounds of sirvinta, gradually amassed huge personal collections of seed stock. Bonsall's biggest collection is seed potatoes; at its peak, he had seven hundred varieties. Now, though, the collection is down to about three hundred. Unless he can find a home for the whole collection or several individual stewards willing to continue growing out the many varieties, this important resource will disappear.

Along with fertile soil and good seeds, the hippie-era back-to-the-landers needed tools. This, too, was tougher for them than for earlier back-to-the-landers. As farms grew bigger, so, too, did farm machinery. Their small plots were ill-suited to machines made for industrial-scale farms and would have been prohibitively expensive for most folks. Plus, many homesteaders rejected tools run on fossil fuels. Whether using tools powered by humans, horses, or diesel, growers had to be able to prepare beds, measure out and plant seeds, and weed, harvest, and clean their crops. Some tools could be found in the *Whole Earth Catalog*, which billed itself as a resource for "appropriate technology"—small-scaled, energy-efficient, and environmentally kind tools.

Other appropriate technologies were whatever one had at hand that could be repurposed. I use a broadfork to make evenly spaced seed holes for certain crops, for example, because I still don't have a dibble. Eliot Coleman recalled that when he started farming, the only piece of farm equipment he had was a four-horse power Troy-Bilt rototiller—so it became his go-to tool for lots of tasks it hadn't been designed for. But even the multitalented rototiller couldn't cut through all the roots in the fields he had to stump. A neighbor was using an old prairie breaker plow to hold up his mailbox, so Coleman traded him a new post for the plow. Chaining it to his jeep, he ripped through the roots. Jerry-rigging of this sort was common on the farms and home gardens of those back-to-the-landers, as they sought to limit their consumption. Making do was a broadly shared value.

Recognizing what an impediment it was not to have right-sized equipment, Coleman later joined forces with Jack Algiere, from Stone Barns Center for Food and Agriculture, and Adam Lemieux, a longtime employee at Johnny's Selected Seeds, and a few others. Beginning in 2012, they met annually to (re-)invent what they've dubbed "slow tools"—human-scaled, ergonomically designed tools for small farms. Among their innovations are a tiller powered by an electric drill; a double wheel hoe; a four-row pin-point seeder; soil blockers; crop protectors; hoop benders; and dozens of other items. These non-diesel-powered tools enable farmers to be, as Algiere put it, "a little more intimate by getting us back into the soil." They also relieve the current generation from having to come up with all those farm hacks on their own.[21]

CONSPIRATORIAL ABUNDANCE

In 1968, inspired by the Diggers' Free Food, households and collectives in the Bay Area began pooling their grocery money to buy grains and beans in bulk and to buy fresh organic produce, tofu, cheeses, and other items supermarkets didn't stock. Among the earliest was the Haight-Ashbury Food Conspiracy. John Carter, an early member, explained that calling it a "food conspiracy" was meant satirically: the "rhetoric of the time had us all as communist conspirators against the state, the war and public morality. We tended to take the charge as a compliment." Indeed, a food conspiracy was about far more than food. Carter called the Haight-Ashbury Food Conspiracy "a housing resource, a study guide, a tool for organizing our anti-war protests. We'd use it to coordinate who was bringing the cloths soaked in vinegar to use when we got teargassed." Abbie Hoffman described a food conspiracy as "a ready-made bridge for developing alliances with blacks, Puerto Ricans, Chicanos, and other groups fighting our common oppressor on a community level."[22]

Conspiracies also influenced non-culinary behavior in less overtly political ways. Lois Wickstrom, a member of a Berkeley Food Conspiracy and author of *The Food Conspiracy Cookbook*, wrote, "The conspiracy has changed us. We are all more conscious of how we spend our money and where we spend it." Wickstrom went on to note, "We had a labor-gift exchange in which all skills are filed so that those who could plumb or do wiring or replace windows could contact those who could sew, cook, garden, or help move furniture. Buying food with your neighbors is just a start. Sharing food with your neighbors leads to a community." By 1975, more than five thousand—and perhaps as many as ten thousand—conspiracies were operating nationwide in both urban and rural areas.[23]

Two of the back-to-the-landers I interviewed, along with another friend, founded a food conspiracy in 1973. John Bly, Craig Wilgus, and George Baggett started the Fog Horn Co-op, and for almost a decade, members made monthly orders. The co-op rented a truck and someone—usually Craig—drove two hundred miles to Boston, where he bought goods in bulk from Erewhon and from the wholesalers at Chelsea Market, the mammoth terminal market that had supplanted Faneuil Hall in 1968. Craig loaded up on "cheese, grains, et cetera, everything but produce. We'd bring it back up and people would arrive at the Ocean View Grange, and we'd distribute it to individuals who would take it home and distribute it to their neighborhoods. We were running loads of five thousand to six thousand dollars a month and filling up a twenty-four-foot U-Haul truck right to the brim."

Area residents still mention those monthly gatherings to break down co-op orders, and when they do, they often sound wistful. Some recall that break-down was usually followed by a potluck supper, others that distributing food to neighbors gave them a chance to socialize and to check on folks who were older or ailing. Though food conspiracies were countercultural, the Fog Horn Co-op members weren't all hip newcomers. Some local families, folks who could trace their roots on the peninsula back two hundred years, also joined. At Fog Horn's membership peak, almost a quarter of the households in town were participating.

WIDENING THE CIRCLES: FOOD CO-OPERATIVES

FOR BACK-TO-THE-LANDERS SIMPLY SEEKING a degree of self-sufficiency, a big garden and a food conspiracy could provide their sustenance. But many who went back to the land in the late 1960s and the 1970s were committed to organics as part of a strategy to address environmental degradation and to weaken the dominance of companies such as Monsanto and Dow Chemical, which made not only agricultural chemicals but also napalm and Agent Orange. Such folks, as Michael Pollan wrote, "sought to establish not just an alternative mode of production (the chemical-free farms), but an alternative system of distribution (the anticapitalist food co-ops), and even an alternative mode of consumption (the 'countercuisine'). These were the three struts on which organic's revolutionary program stood." Transitioning from food conspiracies to storefront cooperatives was a logical next step in strengthening the alternative distribution system.[24]

Just as the food conspiracy members shared the costs for renting vans and other embedded expenses, food cooperative members divvied up the expenses of owning and running the store. Members also did some of the work to run it and made business decisions democratically. By adhering to those criteria, they crafted a kind of capitalism-sans-capital. Money circulated, but the supply chain was short, and the owners and the customers were frequently the same people. As a result, few middlemen profited from the exchanges, and capital didn't accumulate. If a co-op did have an annual profit, members voted on how to use the money. For people seeking more equable forms of exchange than capitalism, co-ops were—and still are—appealing.

During the 1970s, between five thousand and ten thousand consumer co-ops were established in the United States. Although exactly how many of those were food co-ops is unclear, in 1979 some three thousand food co-ops and buying clubs existed. Their allure was strong. Foremost, their focus on selling raw, natural, unprocessed (or minimally processed) foods resonated at a moment when Americans were, once again, particularly concerned about food safety.[25]

From the late sixties into the early eighties, the Food and Drug Administration issued a series of warnings and outright bans on common food additives they'd discovered were dangerous, heightening the anxiety DDT had triggered. Warnings about cyclamates and MSG were announced in 1969; concern about mercury in canned tuna was raised in 1970; the use of the disinfectant hexachlorophene (which was in many items, including baby powder and bath soaps) was curtailed significantly in 1971; the synthetic hormone diethylstilbestrol (DES) was banned in 1972; three food colorings were banned in 1973, saccharin warnings appeared in 1977 (and were elevated to a ban in 1981); and maximum nitrate levels in water were established in 1978. As the grocery aisles became edible minefields, shoppers demanded increased regulation and better labeling. In 1978, several federal agencies held public meetings about the safety of foods, additives, nutritional claims on labels, and the like. However, no new regulations regarding standards or labels were introduced.[26]

Unprocessed foods offered a seemingly safer alternative. No doubt another source of attraction was their prices, which were often lower than those at supermarkets. Low prices appeal at any moment, but they really appeal when there's double-digit inflation. In 1973, food and beverage prices rose 13.29 percent, prompting a weeklong nationwide boycott of meat. And toward the end of the seventies, food prices had two more years of double-digit inflation. Co-ops began to extend beyond the hippie enclaves into new communities. Even as they did so, the co-ops retained their core attributes: They served as community centers, they were co-owned, and they were welcoming places to listen to lectures, exchange recipes, and find bulletin boards papered thick with relevant information.[27]

The Good Tern, the co-op I belong to now, provides an important outlet to farmers and artisanal food makers in the region. In addition to selling locally grown fruits and vegetables, it features locally produced meats, yogurts, cheeses, breads, preserves, soaps, tinctures, salves, and lotions. For me, buying food there is a way to support a local economy that includes people I've met. When I eat Greek yogurt or ground beef from Maine Milkhouse, I can recall sitting in Andy and Caitlin's kitchen, hearing about their path to the farm. Beets and carrots from Ironwood Farm remind me not only of Justin and Nell, but also of the horses who power their farm. Buying greens grown at Turner Farm out on North Haven keeps me mindful of how fortunate we are to have US Representative Chellie Pingree advocating for farmers at the federal level. Bringing home a jar of Turtle Rock Farm's amazing preserves might conjure up the heady scents of rhubarb and Earl Grey tea that bathed Jenn Legnini's commercial kitchen the afternoon I visited. Co-ops offer a visceral sense of being in community, with food as the connective tissue.

For three years, I was a working member of the Good Tern. Like most other co-ops, it offers tiered discounts: The amount depends on whether you're a working or non-working member, a person or a business (corporations are *not* persons at the Good Tern). Some 20 percent of member-owners are working members at any given time, tackling tasks that require little training, like stocking shelves, re-bagging bulk items, and cleaning. My shifts usually involved breaking down large quantities of organic nuts or dried fruits or the like into household quantities or refilling the bulk herb and spice and tea canisters, an exercise in olfactory overload I loved. Volunteering almost always meant a pleasantly repetitive, meditative afternoon spent in the service of collective sufficiency.

MARKETING FARMERS' MARKETS

UNLIKE THE EARLIER GENERATIONS of back-to-the-landers, the Baby Boomers came of age with little access to farmers' markets. From the colonial period into the early decades of the twentieth century, even urbanites could buy food directly from farmers. Some was sold door-to-door from pushcarts, some in open-air markets, some in indoor public markets. As America's transportation infrastructure improved and as farm sizes grew, those farmers' markets declined. Faneuil Hall, in Boston, the oldest indoor public market, ceased operating as a farmers' market in 1968. It was replaced by the wholesalers at Chelsea Market; north of the city, the new locale could be served by trains as well as trucks. Nationwide, only about a hundred farmers' markets remained.

My early memories of farmers' markets are far from idyllic. In the early seventies, on trips into Boston to the Children's Museum or to visit my grandparents, my parents sometimes stopped at the outdoor Haymarket for fruit and vegetables. We kids were warned to *hold hands and stay together!* an exhortation easily obeyed as the crush of people pushing and shoving carried us with it; we couldn't have dispersed if we'd wanted to. We were too short to see the wares but the perfect heights to see that the sidewalk was thick with trampled, gooey produce—reeking in some places, merely a slippery mess in others. Cardboard boxes, broken pallets, and gone-by produce were piled high beside or behind each vendor's stall. And every thirty feet or so we'd pass an ill-lit doorway to a shop promising "bargain basement" prices on seafood or meat, the jamb as often as not framing a man in a bloodstained apron.

Much later I'd learn that the pushcart operators at Haymarket weren't farmers. Farmers had been supplanted by small-scale sellers who bought ripe (sometimes overripe) fruits and vegetables at deep discounts at Chelsea Market on Thursdays. From sunup until sundown on Fridays and Saturdays, they'd sell them at Haymarket. Items were priced well below supermarket prices, often a quarter or a third as much. Midafternoon on Saturdays, the vendors would start cutting deals to get rid of whatever they had left. Bargain seekers shopped then, haggling to get prices down even more because they knew a bag of apples likely contained a couple rotten ones or that one or two of the dozen peaches would be moldy.

Some of the remaining markets in the United States were in much better shape than Haymarket was. A few were even meeting their original aim of connecting farmers to eaters. Reading Terminal Market, in Philadelphia, was going strong, as was Seattle's Pike Place Market. The market in Lansing, Michigan, was faring less well, but when city officials had to vote whether to keep or demolish it, they voted to keep it because it provided "one of the few contacts for our young people between urban and farm activities." The city of Rochester, New York, was running a farmers' market in the hope that it would attract people who'd moved to the suburbs to come back downtown to shop.[28]

The Rochester market caught the attention of Barry Benepe, who was an urban planner in New York City during the 1970s. Benepe was dismayed by the lack of good produce available there and by the fact that farmers in New York and New Jersey were feeling forced, financially, to sell their land to developers. Offering a variant on Thoreau's berry complaint, Benepe lamented, "The quality of food in our stores was terrible! In August when peaches are ripening in Long Island, they weren't appearing in

New York. They were hard and green and from California or somewhere else. . . . everything was wrapped in shrink wrap. We didn't have a sense of real food." To solve both problems, he and a colleague sought to "link the sense of a farm economy with a concept of nutrition and enjoyment of eating food," by creating farmers' markets, which they called Greenmarkets. In the summer of 1976, the first Greenmarket opened at the corner of 59th Street and Second Avenue, with just four vegetable farmers, one egg seller, and two sellers of house plants. Within a month, all eighteen of its stalls were occupied; by summer's end, two more Greenmarkets opened. Today, New York has more than fifty.[29]

Farmers' markets nationwide experienced a resurgence, growing from a hundred in 1960 to approximately six hundred in 1976. Benepe's goal was to keep existing local farmers on their land, but Robert Sommer makes clear in *Farmers Markets in America: A Renaissance* that many of those who participated in farmers' markets were back-to-the-landers, or, as he called them, "children of the ecology movement" who gave their farms "funky names like Good Humus, Mr. Cleanjeans, Chicken Rick, and The Good Earth Cooperative." When Robert Houriet suggested that NOFA seed Vermont with farmers' markets, he was tapping into the zeitgeist in a way well suited to the region. By the end of the decade, thirteen Vermont farmers' markets were up and running.[30]

The success of farmers' markets was fueled, in part, by the same anxiety that led people to demand better labeling with respect to pesticides and additives; they wanted safe food. Also, during the seventies, food at farmers' markets was often considerably less expensive than it was at grocery stores; comparing prices at fifteen California farmers' markets and nearby supermarkets, researchers found prices to be 34 percent lower at the farmers' markets. Rising transportation costs and the growing number of intermediaries in the industrial agriculture system were driving up costs, so much so that by the end of the decade, only 30 percent of a consumer's food dollars went to the farmer.[31]

In 1976, the farmers' market movement gained tremendous momentum with the passage of the Farmer-to-Consumer Direct Marketing Act. While the growth between 1960 and 1976, from one hundred to six hundred markets, was impressive, numbers skyrocketed after the Direct Marketing Act passed. By 2019, the USDA counted more than 8,100 farmers' markets nationwide. Most of a market's vendors—many of them members of the current generation of young farmers—work within fifty miles of it. That factor is part of their appeal. By fostering local resilience and independence, farmers' markets offer a measure of food security in an uncertain world. As Sommer observed, "Local producers and local markets are social buffers in times of crisis," as local food is still available if there's an oil shortage or a transportation strike.

Such strikes hit shelves hard a few times in the seventies. In 1974, an eleven-day nationwide truckers strike left the shelves half empty in grocery stores across the country. During another, in 1979, "about 80 rigs parked bumper-to-bumper to shut down the New England Produce Center" and had to be dispersed by police. Three-quarters of the meat shipments sent to the East Coast never reached their destination. More recently, as the global supply chain failed again and again during the COVID-19 pandemic, many Americans rediscovered local farms, and began to sense the crucial role they can play in maintaining food access.[32]

U-PICK FARMS AND CSAS

FOR MANY NEW YOUNG farmers, a farm stand and a farmers' market are the primary ways to get their wares to customers. Others tap into more traditional retail methods: They have standing orders at regional co-ops, work with wholesalers, or participate in food-to-institution programs with schools or hospitals. In addition to these distribution methods is Community Supported Agriculture, a variation on the Clientele Membership Club model that Booker T. Whatley proposed in the 1970s.[33]

A horticulturalist at Tuskegee Institute, Dr. Whatley said in an interview in 1982 that he and other professional agronomists had "made life hard on the small farmers by recommending that they grow exactly what the big boys produce." He'd come to realize that because commodity crops have such a low return, the small-scale farmer "works hard and just about starves to death." Instead, he proposed that small-scale farmers grow diversified specialty crops. But he worried that farmers' markets and food co-ops might not provide enough customers, so he suggested a plan to leverage urbanites' interest in quality food. Farmers should create Clientele Membership Clubs, and club members would come to the farms and harvest their own food. He also recommended that farmers lease their trees for fifty dollars each. Clients would pay at the start of the season and then, come fall, gather nuts or fruit from "their" trees. Quick to deflect skepticism, Dr. Whatley acknowledged "some folks might think that a person wouldn't pay $50 to gather the nuts from one pecan tree, but a lot of them will! It really happens."[34]

Dr. Whatley's small-farm plan isn't as well known today as either U-Pick farms or Community Supported Agriculture (CSA), although both are included in his model. CSAs surged in popularity in the United States during the 2000s and 2010s. In 2015, the most recent year for which the USDA has collected data, 7,398 farms offered CSAs. Typically, CSA members enroll and pay the farmer at the beginning of the season and then receive food in weekly installments—though unlike in Dr. Whatley's plan, CSAs don't usually allow clients to harvest their own food. CSAs help the farmer by providing an infusion of cash when it helps most, as she's gearing up for the season. And the eater has fresh, local food whose source she gets to know well. Their reciprocal commitment strengthens interpersonal bonds and improves the local economy. And CSA members take on some of the risk inherent to farming, as the content of their boxes is contingent on the weather. Although sharing risk isn't as appealing as getting to know the farmer and enjoying local foods, it does remind eaters what's involved in getting food to plates.

Because we have such a large garden, I thought the CSA trend would pass us by. But in 2008, a group of fishermen on our peninsula began Port Clyde Fresh Catch, the first Community Supported Fishery in the nation. As one would in a CSA, we paid for a season up front, then for the next ten weeks, we received a portion of whatever the groundfishing fleet caught. On Saturday mornings, I drove to the nearest pickup point, a gravel driveway a mile down the road, bringing extra plastic grocery sacks to triple bag my catch. Gary Libby, one of the fishermen, opened the cooler resting on the flatbed of his truck and pulled out something that glistened even in his spruce-shaded yard. Often it was a haddock, or a pollock, which looks like the Platonic ideal of fish; twice it was a prehistoric-looking monkfish, its appearance entirely at odds with its deliciousness. Once, he handed me a bag containing half a dozen flounder, fish so flat I completely mangled the first two before making a few somewhat (okay, barely) acceptable fillets.

The CSF enhanced my appreciation of eating in season. The ocean is as seasonal as the land, a fact I now keep in mind if I crave shrimp in July. And I discovered the ocean to be home to lots of tasty species that were never in, or have fallen out of, culinary fashion—like the redfish I'd never heard of, much less eaten, until then. The CSF lasted just a couple of years before it morphed into a fish market; convincing people to gut fish each week is way more difficult than getting them to try kohlrabi. But even veggie CSAs are facing challenges.

Diane Einseidler, who apprenticed at Eliot Coleman's Four Season Farm and now works at Harlem Valley Homestead, suspects that both farmers' markets and CSAs have peaked. "Farmers' markets aren't very functional for the farmer," she told me, and are "limited in who they reach," and CSAs demand a level of commitment sure to limit their growth. She went on: "There's a lot of talk now about the institutional CSA model and in permanent farmers' markets that you don't take down and set up, places like the Boston Public Market. The way we see it, the CSA model was a way to reach more people at a specific point in time in the movement, but it's reached its limit." Doug Fox, who teaches Sustainable Agriculture at Unity College, agrees. He sees additional options emerging to solve these economic problems, including food hubs and other collaborations. He hopes—and believes—the result will be "a new economic system arising within capitalism."[35]

"ALICE WATERS ENNOBLED US"

As discussed earlier, school lunch programs helped to Americanize children's diets by feeding them "low-toned" foods during World War II. In the war's wake, Americans zealously chose frozen foods over fresh and ate many a meal in the newly created fast-food restaurants. All three of those dietary shifts reduced the breadth of flavors in the mainstream American diet. Likewise, the increase in the size of farms after World War II, thanks to the introduction of synthetic fertilizers, led to many changes. The surfeit of corn, for example, prompted the development of high fructose corn syrup (HFCS). In turn, HFCS altered the flavor profiles of the foods in which it replaced sugars.

The greater size of their farms also led farmers to want more-uniform produce so it could be picked and packaged mechanically. However, as breeders selected for uniformity, or for other traits large-scale growers value, such as a long shelf life or an unblemished appearance, they often had to forsake flavor. Thus, many large-scale industrially grown vegetables and fruits have gradually become less flavorful than were their antecedents. And with palates being shaped by processed foods and less piquant fruits and vegetables, people increasingly reported that sharp or acerbic or bitter flavors were unappealing.

To accommodate changing flavor preferences, plant breeders began editing out the genes controlling the production of the chemicals that generate those strong flavors. Unfortunately, doing so doesn't simply reduce a plant's flavor profile; it also makes the plant less nutritious. The omitted chemicals, which are part of the plants' immune system, have health benefits both for the plants and for those who eat the plants. As this cascade of effects began with the introduction of synthetic fertilizers, the chef Dan Barber described them as "disastrous for the flavor of food," while the epidemiologist Adam Drewnowski worries about the implications for human health that come with breeding out those beneficial chemicals.[36]

Whereas industrial practices have led to many foods being less flavorful and less nutritious, the practices the seventies-era back-to-the-landers promoted have the capacity to increase foods' flavors. Growing food in nutrient-rich soil, choosing cultivars for flavor rather than for some other attribute, and selling produce soon after it's been harvested all enhance flavor. The farmers and organic growers certainly believed their produce and artisanal fare tasted great—but convincing the public required others to be equally enthusiastic. Chefs and restaurateurs, especially Alice Waters, deserve much of the credit for that. Eliot Coleman put it well: "Alice Waters ennobled us. She led the drive that's now affecting all the chefs to say 'Hey, we can't do anything good unless we're getting good ingredients. And good ingredients come from small farms.'" And from small artisanal producers, like Laura Chenel, whose goat cheese Waters so loved.

Alice Waters came of age politically in Berkeley, California, and gastronomically in France. In her memoir, *Coming to My Senses*, she wrote that she'd always been an activist and a sensualist. She recalled listening to Mario Savio at Sproul Plaza in

December 1964 and feeling so inspired that she started to believe her generation could change the world. "We didn't just think we could stop the war," she wrote, "we thought we could create an entirely new sort of society." Not long afterward, Waters left Berkeley for a semester abroad in France, where she concentrated far more on food than on school. Leisurely afternoons in cafés, lunches at upscale restaurants, and dinners at unprepossessing spots with just half a dozen tables filled her time. She fell in love with "market cooking," organizing meals around whatever looked perfect at the neighborhood's outdoor market on any particular morning.[37]

Waters described a life-changing lunch at a tiny restaurant in Brittany. "The menu on that day," she wrote, "was cured ham and melon, whole trout with slivered almonds in browned butter, and a raspberry tart." Thinking back on it, Waters realized the simple foods were so good "because the trout probably came from the stream we could see out the window, the melons came from the garden, and the owners likely made their own *jambon*. It was one of those perfect little meals. Years later when I opened Chez Panisse, I looked back on that experience as a blueprint."[38]

Returning to Berkeley to finish school, she began a cooking column for one of the alternative newspapers. In a puckish nod to Arlo Guthrie's song about Alice Brock, she called it "Alice's Restaurant." But Waters's ideas couldn't have been farther from those of her East Coast namesake. Whereas Alice Brock blithely assured readers that adding soy sauce makes a dish Chinese, Alice Waters wanted nothing to do "with the hippies' style of cooking: a jumble of chopped vegetables tossed together with pasta—throw in a few bamboo shoots, and call it a Chinese meal."[39]

Nevertheless, when Waters opened Chez Panisse, in 1971, she had to buy ingredients in the same places the hippies were getting theirs—co-ops, ethnic markets, and local wholesalers. Even those supermarket alternatives weren't carrying a lot of the ingredients she sought. Many items considered basic today—like extra-virgin olive oil, nut oils, snap peas, yellow peppers, fresh pasta, and fresh herbs—were simply not available. To source more interesting options than were commercially available, friends and employees foraged, creating what one of Waters's friends called the "hunter-gatherer culture" of Chez Panisse: "We gathered watercress from streams, picked nasturtiums and fennel from roadsides, and gathered blackberries from the Santa Fe tracks in Berkeley. We took herbs from the gardens of friends. We also relied on friends with rural connections. The mother of one of our cooks planted *fraises de bois* for us, and Lindsey got her father to grow the perfect fruit she wanted." Friends would augment what Waters found, bringing "wild mushrooms from Mount Tamalpais, huckleberries from Point Reyes, a better egg from some old farmer's lost race of chickens, an incomparable plum from somebody's brother's backyard."[40]

It's almost impossible to overestimate how much her simple-sounding recognition that great meals depend on great ingredients ran counter to the food culture then, or how much it transformed that culture. In *Ten Restaurants That Changed America*, Danny Meyer and Paul Freedman include Chez Panisse and list the "most influential" of Alice Waters's ideas: "Primary ingredients must be of high quality; quality must be defined in terms of freshness and naturalness; freshness and naturalness are to be thought of in terms of seasonality, location, and small-scale, non-industrial agricultural practices." Noting that "these principles are now taken for granted," they write that when Chez Panisse was getting started, "they were not only new but peculiar, virtually unheard of."[41]

To ensure the restaurant's ongoing access to quality ingredients, Alice Waters and the employees at Chez Panisse cultivated relationships with suppliers as carefully as those suppliers cultivated their produce and goods. Working with farmers like Bob Cannard, of Green String Farm, she learned to trust their knowledge; "We just started buying whatever they told us to buy," she said. "[That] changed the restaurant dramatically, to have an understanding of the intellectual work required, to understand the nourishment" organic farmers pack into produce. Chez Panisse has sourced herbs and vegetables from Bob Cannard for more than thirty years. His farming methods are themselves rooted in the belief that "good relationships" are crucial. Seeing that cooperation among species is the norm in nature, Cannard mimics this principle as he replenishes depleted pieces of land.[42]

Alan Tangren, whose job for a while was to ensure the ingredients were stellar, considered those relationships with farmers mutual. He described the food at Chez Panisse as "a real collaboration between the farmers and the restaurant. They brought us the best things they had and we used them." In part as a result of this consistent access to marvelous raw ingredients, the fare at Chez Panisse became simpler over time. The titles of Waters's most recent cookbooks, *The Art of Simple Food* (2007), *The Art of Simple Food II* (2013), and *My Pantry: Homemade Ingredients That Make Simple Meals Your Own* (2015), announce this emphasis. The recipes and preparation styles are, indeed, less complex than those in her early cookbooks.[43]

Waters's commitment to aesthetics and to flavorful, often expensive food may seem out of kilter with the preferences of back-to-the-landers. However, like them, she regards cooking and eating in ethical and political terms. She stresses how important food can be to community—that who one eats with and what one eats reflect values. By the late 1980s, Waters was using her celebrity to advocate for sustainable agriculture, food justice, and food-access programs. She imagined transforming empty lots into urban gardens (the idea wasn't in vogue at the time). When she decried conditions at a nearby middle school, its principal invited her to visit; she proposed transforming an acre of the school's asphalt lot into a teaching—and eating—garden. With help from the school's staff, donors, and other supporters, the

Martin Luther King Jr. Middle School became home to the first Edible Schoolyard. Today, more than fifty-five hundred edible schoolyards help children grow and learn about food while enhancing their food security.

Waters's commitment to clean, whole foods was not inspired by the abstemious Alcott or the disciplined Nearing. Her counterculture hero was William Morris, the Arts and Crafts movement front man whose ideas inspired Arden's founders. Waters wrote that she "adore[s] William Morris," especially his belief that "art must begin at home." In her memoir, she paraphrased his central design principle—"Have nothing in your house that you do not believe to be useful, or believe to be beautiful." Like Morris, Waters values beauty and artisanal craftsmanship. She sees creating and celebrating them as socially radical, a way to undercut the alienating effects of an often shoddy industrial culture. Even if she hadn't name-checked Morris, her admiration is readily evident: All of Chez Panisse's early branding—its posters, announcements, and cookbook covers—employed Arts and Crafts–style fonts and graphics.[44]

LAND AVAILABILITY

By the time Alice Waters started looking for small farms from which to buy ingredients, finding any close to San Francisco was difficult. When Chez Panisse opened, 80 percent of the state's agricultural land was in farms of at least five hundred acres. And even though real-estate prices in the Bay Area were lower in the pre–Silicon Valley days, they weren't low enough to make land affordable for small-scale farming. As was happening on the periphery of cities across the nation, the price for open land in the Bay Area rose rapidly after World War II, when urban sprawl transformed undeveloped spaces into shopping malls, subdivisions, gas stations, big-box stores, motels, and fast-food restaurants.

The building boom was part of a federal effort to stave off a serious recession or depression. During the war, home construction had dwindled. Where would the six and a half million returning GIs live? And what would they and the eight million people who'd held war-related jobs do to earn money? To promote job creation, home-buying, and a stable economy, the Fed kept interest rates incredibly low. Manufacturers expanded, builders built, and consumers borrowed money to purchase new homes, new cars, new anything. With a hearty appetite for *stuff* after years of doing without, folks didn't want to wait to buy. To help appease pent-up demand, the government loosened credit regulations, making it easy to get credit cards, store credit, personal loans, and payday loans.

In 1954, commercial sprawl was further encouraged by a small change in the tax codes: The IRS revised the formulas for determining how investors calculated depreciation on new commercial construction, and suddenly real-estate development was a profitable tax shelter. As the economist Thomas Hanchett explained, "Savvy investors quickly discovered that they could build a structure, claim 'losses' for several years while enjoying tax-free income, then sell the project for more than they had originally invested." If that sounds like a way for developers to make lots of money with virtually zero economic risk, that's because it is.[45]

The result? "All over the United States, shopping plazas sprouted like well-fertilized weeds." In 1953, the year before the tax-code change, the United States had two hundred and five shopping plazas; five years later, the number had more than tripled, to seven hundred and thirty. Instead of being sited within suburbs, as they had been before the new formula was implemented, the new "shopping centers began appearing in the cornfields *beyond* the edge of existing development." By 1970, thirteen thousand shopping centers stippled the American landscape, many on prime farmland.[46]

Despite all that development, back-to-the-landers in the 1960s and seventies were able to buy land for what now seems like a song. Eliot Coleman bought his sixty acres in Harborside for two thousand dollars. In Monroe, Maine, about fifteen miles inland, Jay Davis bought one hundred and thirty acres with a house and barns for twenty-one thousand dollars. And in St. George, Craig Wilgus paid just six hundred dollars for ten acres. These low prices were due to the land being far from a city or a big town. But within a decade, even prices in Ultima Thule began to rise.

Whereas the federal government quickly came up with a plan to avoid postwar recession, it struggled to address the sprawl its economic plan engendered. In the early 1970s, Congress almost established a federal land use planning program. The sensible-sounding act had two goals: The first was to "assure that the lands in the Nation are used in ways that create and maintain conditions under which man and nature can exist in productive harmony and under which the environmental, social, economic, and other requirements of present and future generations of Americans can be met." And the second was to "support the establishment by the States of effective land use planning and decision-making processes that assure informed consideration, in advance, of the environmental, social, and economic implications of major decisions as to use of the Nation's land and that provide for public education and involvement." But the bill failed. Absent a federal planning mechanism, land-use policies were set—if they were set at all—at the state and local levels.[47]

Seeing the consequences of rapid sprawl for nearby farms (and craving delicious food) had prompted Barry Benepe to create Greenmarkets, which did help some farmers stay on their land. Other farmers' markets no doubt helped farmers do the same. But between 1960 and 1978, notwithstanding the surge of young homesteaders starting farms, the total number of small farms declined by 40 percent. More than a million farms were lost in fewer than twenty years, the majority from high-quality farmland near dense urban markets.[48]

Like Benepe, many other individuals took it upon themselves to try to save land from being developed. Quite a few used a model established in 1891 but rarely used since—the land trust. The first land trust in the United States, the Trustees of Reservations, in Massachusetts, was the brainchild of Charles Eliot, a Boston brahmin trained as both a landscape architect and a horticulturalist. As Boston rapidly—and fetidly—industrialized, Eliot sought a way to preserve scenic land for the public "just as a Public Library holds books and an Art Museum holds pictures." The trustees acquired their first property in 1892 and have been successfully conserving property ever since. But for decades, few others copied their model; at the end of World War II, the United States had only fifty local land trusts.

Then, in the seventies and eighties, land trusts emerged as a citizen-led work-around to preserve open spaces. With sprawl accelerating and the environmental movement well established, landowners saw value in ceding future development rights to preserve the land. By 1981, the number of land trusts had risen to three hundred and seventy. Today, some seventeen hundred land trusts dot the nation.[49]

Les Hyde was one of the architects of the Georges River Land Trust, which protects land within the river's two-hundred-and-twenty-five-square-mile watershed. Before moving to Maine, Les worked for two years at Cornell University, as the assistant director of the Northeast Regional Center for Rural Development, traveling the region's twelve states to learn how they managed sprawl. "The best farmland was being converted to subdivisions and shopping malls," he lamented to me, and the center wanted to figure out what it could do. He slowly realized that governments weren't willing to take a major role in preserving open spaces and farmland, that people "had to create these private nonprofit land trusts" to do it.

Developers eventually came to the St. George Peninsula. When a realty company began buying riverfront property, Les Hyde knew the moment had arrived. In January 1987, he and a few other folks organized an informational meeting to get a sense of community interest in a land trust. Encouraged by the enthusiastic turnout, the core group established the Georges River Land Trust in just three months. Today, it owns or holds conservation easements on four thousand acres. Nationwide, more than fifty-six million acres are similarly protected by local and regional land trusts.[50]

These efforts to relearn how to farm organically, to create markets for organic and local ingredients, to supply those markets, to work with chefs, and to keep land accessible for small-scale farming were all accomplished by folks often criticized for dropping out. Those involved in promoting healthy foods were doing something else, something with far-reaching effects. In fact, David Kamp proposed that "the fresh-food movement . . . might well be the counterculture's greatest and most lasting triumph." They made it more possible to eat safely and nutritiously, and they readied the way for the Millennials who've gone back to the land.[51]

⫸ 9 ⫷

FREE THE PEOPLE, FREE THE LAND

WHEN I BEGAN WRITING this book, I thought I had a clear sense of the scope of racial inequity in the US food system, but I discovered it was even more wide ranging than I'd thought. Although that was disheartening, I also learned about scores of new efforts, many of them grassroots projects, taking place around the nation to redress injustice—on farms, in food deserts, among migrant workers, and among political advocates. Unlike many of the earlier justice efforts within the ongoing Good Food Movement, the newer efforts tend to be led by younger, usually BIPOC, farmers and activists.

In the spring of 2018, Soul Fire Farm began to offer trainings called "Uprooting Racism in the Food System." I knew of Soul Fire Farm because Leah Penniman, one of the cofounders, was—is—an inspiring writer and speaker. I was eager to learn from her and the Soul Fire team and applied to attend a training. So many other people also did that they couldn't accommodate everyone; happily, they decided to host a second training a few months later, and both Rob and I were invited to participate then.

So, on an unseasonably cold September morning, seriously under-caffeinated for the work ahead, we arrived at Soul Fire Farm scant minutes before nine, as did almost everyone else coming for the training. We all shuffled down the long driveway carrying notebooks and phones and dishes for a potluck lunch. Jonah Vitale-Wolff, Soul Fire Farm's other cofounder, welcomed us, then pointed us toward extra sweaters and coats and hats before urging everyone to settle in a circle outside. Twenty-five in all, we were a multigenerational, multihued, sexuality- and gender-diverse crowd, some local, others from as far away as California and England. A handful of the participants were farmers; most worked in farm-adjacent or food- and land-justice jobs.

Leah, Jonah, and two alumnae of Soul Fire's farmer immersion program guided us through activities and tough conversations. Sometimes as a large group, sometimes in pairs or trios, we talked about historical injustices in the food system, white supremacy, and structural racism. Some injustices were single events with immediately significant consequences, like President Johnson's rescinding Sherman's Field Order granting freed slaves forty acres of land. Others had impacts that accumulated over time. Except for slavery, arguably the most long-lived of the discriminatory practices BIPOC farmers have faced were carried out by the USDA.

In 1997, Black farmers sued the USDA in a class-action lawsuit, *Pigford v. Glickman*, arguing they'd been treated unfairly for decades by the officials responsible for approving FSA farm loans, farmer assistance requests, and aid in debt restructuring. Preparing to fight the lawsuit, the USDA hired a consulting firm—which found the evidence so thoroughly supported the farmers' claims that the USDA admitted the discrimination and settled the case. The award, ultimately around two billion dollars, is the largest civil rights settlement in US history. A similar lawsuit, *Keepseagle v. Vilsack*, was filed in 1999 on behalf of Native Americans who were denied loans and forced to sell their land, only to have it go to white people who received USDA loans to purchase it. That suit, too, was settled in favor of the plaintiffs.

Our conversation then turned from past acts to present-day issues, such as food apartheid and its health effects on minority populations, the risks farmworkers face from pesticide exposure, and the conditions in meatpacking plants. The Soul Fire team gently but firmly reminded us that once we know these things, if we don't do anything to redress them, we're tacitly perpetuating them. A young beekeeper told us he'd stopped eating sugar when he learned that much sugar production relies on forced labor; he reminded me of Alcott avoiding molasses because it was a product of the triangular slave trade. A little while later, I was again reminded of earlier back-to-the-landers when Leah led us in a call-and-response: her calls to "free the people" answered by ours to "free the land." Like Alcott and Lane, like the single-taxers, like the Diggers, like Lou Gottlieb of Morningstar Ranch, the Soul Fire Farmers are adamant that the land is theirs to steward and to love, but not to own.

Midday, we convened on the edge of a fallow field, attendees standing shoulder-to-shoulder on one side of a snaking garden hose, Leah facing us from the other side. For more than thirty minutes, she read a series of prompts that all began "Cross the line if . . ." She urged us to take a moment to look at who was on either side of the garden hose after each cue, before thanking us as we recongregated on the far side. The edicts started out easy: "Cross the line if you ate breakfast this morning." To my surprise, only Rob and I stayed put. "Cross the line if you traveled more than an hour to get here." Most of us crossed for that one—not a surprise; Grafton is remote, about ten miles from the Massachusetts and Vermont borders. "Cross the line if you had a cherished natural place as a child." Half and half. The prompts became more pointed: "Cross the line if you've worked some place where your native language wasn't spoken," "Cross the line if you've ever been food insecure," "Cross the line if you inherited or expect to inherit money," "Cross the line if you've ever been threatened because of your race," "Cross the line if your family ever owned slaves," "Cross the line if your family members were ever owned." Dozens of prompts reconfigured us again and again, rarely into the same alignments. Very occasionally, race was the sole distinction between people on either side. Far more often it was a combination of traits—an intersection of race with class, perhaps, or with gender or ethnicity or age or physical attributes. A few people in the group seemed to have been granted great ease in the life lottery; most of us had a mishmash of privilege in some respects and a lack thereof in others.

Mid-afternoon, we were urged to think more deeply about privilege, to identify what we had and how to use it to improve the food system. Leah started us off. She is charismatic and attractive, with a dazzling smile that makes her tawny complexion seem to glow. She said darker-skinned people experience more racism, so she regards her light skin as a privilege she brings to her work. I wrote down whiteness as a privilege, then stopped because all the others I could think of arose from that one. But everyone else was still writing, so I added some specifics: being able to talk my way out of speeding tickets, rarely being the only person of my race in a room, never having a stranger ask to touch my hair, getting a mortgage without jumping through a million hoops.

We talked about how to help dismantle racism without letting privilege undermine good intentions. Jonah let us know he's white, of Italian and Jewish heritage. He said he's acutely aware of the privilege that comes with being male and white, so he strives to follow the lead of the Soul Fire Farmers of color. His words called to mind the advocacy slogan "Nothing About Us Without US."

A half hour into our drive home, traffic came to a stand-still on Route 2 in Massachusetts. An accident had the highway backed up for miles, so Rob and I got off at the next exit and began zig-zagging east through the old mill towns that line the Connecticut River. Navigating the unfamiliar roads, I thought about how risky a detour off a highway might be for someone driving while Black, or undocumented, or non–English speaking, and realized that worrying only about how long it'll take to get home can also be a form of a privilege.

"Using one's privilege"—a twenty-first-century expression—is a trait shared by all the utopian back-to-the-landers chronicled here. Thoreau, for example, understood that whatever privilege he enjoyed obligated him to use it to support others. In his essay "Civil Disobedience," he wrote that he could more readily spend a night in jail to stand up for his beliefs than could most of his neighbors: "They cannot spare the protection of the existing government, and they dread the consequences to their property and families of disobedience to it . . . if I deny the authority of the State when it presents its tax-bill, it will soon take and waste all my property, and so harass me and my children without end." Unlike his neighbors, Thoreau had little property to lose and no children to protect. He didn't have to worry about his actions subjecting others to harassment and understood that this comparative ease made acting morally essential. Likewise, Thoreau and others in the first wave of utopian back-to-the-land experiments, in the 1840s, used their voices, bodies, and/or money to advocate for abolition and other social reforms.[1]

Frank Stephens, Will Price, and Joseph Fels, the men who established Arden during the second wave, at the turn of the century, used their professional skills and economic resources to advocate for and create prototypes of what they saw as fairer alternatives to the overwhelming inequality of cities during the Gilded Age.

When Scott Nearing lost access to commercial publishing houses, he made his own so the couple could use their writing talents to spread their beliefs that a simple life is an ethically and politically charged option to living within capitalism.

Likewise, many of those who went back to the land in the late sixties into the seventies had been involved in war protests, civil rights demonstrations, rallies for women's rights, nuclear energy protests, and environmental actions, using their college newspapers, their bodies, and their networks to advocate for greater justice and equality.

Now the fifth wave, Millennial farmers, use the knowledge and systems created by the previous generations to improve soil, produce and distribute nutrient-dense foods, tend livestock humanely, and seek a fair and healthy life for themselves, their workers, their customers, and their communities. When they advocate for social or environmental reforms or educate others about food or farm life, the privilege they draw on is twinned—melding their credibility as farmers with skills likely gained in college for public speaking and critical writing. Some with financial resources have created nonprofit farms that donate their entire harvest to food banks. In working toward a healthier, more egalitarian world, these Millennial farmers, like utopians before them, expand on the original promise of American freedom.

But obstacles abound. Not only must many farmers navigate the challenges posed by systemic racism and individual racism, but almost all must navigate the roadblocks posed by difficulties accessing land, skewed federal policies, and resistance from customers to paying what seem like unduly high prices. If the National Young Farmers Coalition is correct that this generation *must* be the one to transform the dominant US food system, to replace it with one that is not harmful to the planet, not reliant on exploited labor, and not overproducing foods with limited nutritional benefit and limited flavor, then those of us who value their efforts need to join them in dreaming this better world into being. We must use our privilege.

LAND ACCESS DIFFICULTIES

A FEW MONTHS AFTER participating in the "Uprooting Racism" training at Soul Fire Farm, I was again driving west on Route 2, this time to visit Tyler Sage, my teaching assistant who became a farmer. After several years on rented land, he'd recently found a farm to buy in Bernardston, Massachusetts, a rural community near the Vermont border. His new place consists of 140 acres of mostly wooded hillside, along with enough grazable, flattish acreage to accommodate a farmhouse, a barn, a paddock for two horses, and a quirky outbuilding whose original use eludes me.

Tyler had just finished converting the old dairy barn into a space for pigs and wanted to show it to me. As he loped across the ice in muck boots, I tottered awkwardly in shoes, wishing I'd thought to bring wellies. Once inside the barn, I needed a minute to adjust to the comparative darkness—and longer to get used to the smell. Shards of sunlight filtered through gaps between the old planks, augmenting the light from fixtures Tyler had installed. Each of the newly partitioned stalls had snout-controlled feeders and sets of water nipples so the pigs could mostly feed themselves.

Hardy as they are, pigs don't do well in the cold, so Tyler was using deep-litter bedding to keep them comfortable for the winter, laying down several inches of straw, then adding a new layer every few days to cover the one saturated with pig waste. Come spring, he'll muck out the dense mass and let it finish turning into nutrient-dense compost. Deep-litter bedding is gaining popularity on small-scale farms because it provides animals with a place to root and, as the materials compost, the bedding gives off heat, helps manage pathogens, and—supposedly—reduces the ammonia smell from the pigs' urine (I'm not entirely convinced of that one).

The animals were grouped by size, so walking from one stall to the next was like fast-forwarding through the six months it takes for pigs to grow from a farrow of adorable three-pound piglets to a herd of vaguely intimidating ready-for-slaughter two-hundred-and-fifty-pounders. The newest litters were so tightly packed together, as they clasped their sows' teats, that I could only differentiate the hungry mass into individual animals if some of them had distinct markings. When we paused in front of a stall of two-month-olds, Tyler introduced some of the pigs by name, and continued to do so as we made our way around the barn.

After I met the pigs, we headed to the paddock to say hi to the horses, Shorty and Rosa. I'd seen photos but hadn't met them in real life. I was instantly smitten with Rosa, a beautiful Suffolk with a placid temperament and frank gaze. She let me scratch her muzzle and chin groove while Tyler clowned around with Shorty. Tyler said he considers the horses coequal team members, which I loved; it reminded me of Caitlin Frame saying she considered the cows and grass her partners at the Maine Milkhouse.

I'm ridiculously pleased for Tyler. He'd been working to get to that point for eight years. Because he went to art school, he had to learn pretty much everything about being a farmer afterward. His first apprenticeship was at Moon in the Pond Farm, which had every farm animal he could think of. After that, he headed to Blue Hill Farm, co-owned by the renowned chef and food activist Dan Barber, and eventually finished his apprenticeships with one on a horse-powered farm to learn to work with draft animals. Somewhere along the way, he trained as a butcher. He sees being able to butcher as part of fully relating to his animals and his food, something he wants to be able to do in general. "I see it replicated in other instances in my life," he told me. "If I have to build a thing, I don't really want to go to Home Depot and buy a bunch of two-by-fours and put it together. I want to harness up my horse; I want to go into the woods; I want to cut down some trees, stack them up, get a sawmill, mill them up into lumber, and then build." Like other back-to-the-landers who cherish self-sufficiency, Tyler finds it empowering to understand all the processes involved in meeting his needs.

But self-sufficiency doesn't mean going it alone. Not only did he take on multiple apprenticeships, he's also gotten educational and business support from farm-centered nonprofits. Once he was ready to start his own farm, Tyler got in touch with the Carrot Project, a Massachusetts-based nonprofit that helps new farmers get loans and develop financial and business plans. When it was time for him to find his own place, he got assistance from the American Farmland Trust, which was founded in 1980 to do exactly that. Whereas conventional land trusts often try to maintain the land unchanged over time, agricultural trusts allow the land to be used in different ways, as long as it remains farmed.

Being able to afford farmland is, of course, also a form of privilege. So much so that respondents to a survey by the National Young Farmers Coalition named access to land as their number-one challenge. Like Tyler, most of these young farmers start out on rented or borrowed land. Even those who can afford to buy land, or who are able to obtain loans, often struggle to find good land in proximity to a strong market. Tyler is in a good place, the Pioneer Valley, where the farm scene is robust. But even with help from the AFT, he needed a nontraditional mortgage loan, and had applied for one from the Farm Service Agency (FSA).

As we talked that day, Tyler mentioned the loan hadn't yet been approved; he was worried that if the government shut down, as then President Trump was threatening, his chance to get a loan would be undermined. I assured him the threats were all bluster, that no president would close the government the week before Christmas. Then ate my words a few days later.

When the shutdown began, Secretary of Agriculture Sonny Perdue assured the public the USDA would continue "safeguarding

life and property through the critical services we provide." Among
the services he deemed critical were those involved in inspecting
foods, a few international trade programs, and the divisions that
fulfill the Wild Horses and Burros Statute of 1971. Many, many
other programs went on hiatus or got scaled back. Nutritional-
assistance programs were funded only through January or only
with whatever commodities they already had on hand. And just
as Tyler had feared, loans and grants related to rural development
were put on hold. In fact, after the first week of the shutdown, the
hold was extended to all farm loans and payments.[2]

Five weeks after it began, the longest shutdown in US history
ended. The USDA was allowed to fully reopen and help farmers
with all their crucial midwinter tasks—including approving Tyler's loan.

THE USDA AND THE NEED TO
TRANSFORM US AGRICULTURE

THAT SHUTDOWN THREW INTO relief the range of issues, and
disparate priorities, the USDA manages. The notion that maintaining nutritional-assistance programs was less critical than fulfilling the Wild Horses and Burros Statute didn't sit well with
me. Similarly, in 2020, when the COVID-19 pandemic derailed
everything, the allocations of governmental aid cast fresh light
on priorities. Under the 2020 Coronavirus Food Assistance Program, the USDA administered some twenty-six million dollars
that went to farmers. That sounds like a godsend—and for some,
it was. The top 1 percent of recipients got 21 percent of the money, with an average payment of $342,752 each. The bottom 80
percent, on the other hand, *collectively* received slightly more, 23
percent, with average individual payments of just $4,637. Nine-
ty-nine percent of the Coronavirus Food Assistance money went
to white farmers; 0.1% percent went to Black farmers.

In 2021, during the Biden administration, the American Rescue Plan Act directed the USDA to forgive more than four billion dollars' worth of loans for approximately sixteen thousand BIPOC farmers. I don't know the thinking behind that plan, but I imagine it was partly to avoid a repeat of the earlier, lopsided aid allocation. However, a year and thirteen lawsuits later, the loan-forgiveness program has yet to start. Three courts have already issued injunctions on the behalf of plaintiffs who argue the program is unconstitutional because it discriminates based on a farmer's race. Most of the legal commentators who've weighed in now doubt that it will be implemented.

As we've seen, the USDA has prioritized larger farms over smaller ones and white-owned farms over Black-owned farms since its inception. This imbalance has hurt many aspiring farmers—from Thoreau's neighbors in Concord to Black farmers after Reconstruction, to California's small-scale farmers, to those forced off their land during the post-WWII development binge, to small-scale organic dairy farmers now. Not every secretary of agriculture adhered to the bigger-is-better ethos, of course. A few have tried to make American agriculture more amenable to smaller farms and to nonindustrial methods. But corporate and political opposition are so entrenched that these efforts haven't gained much traction.

In the late 1970s, in response to concerns about a "decrease in energy efficiency" of industrial agriculture and "a growing concern about the adverse effects of intensive production of cash grains and about the extensive, and sometimes excessive, use of chemical fertilizers and pesticides," then Secretary of Agriculture Bob Bergland tasked a USDA research team with figuring out if farmers could grow enough food and fiber organically to meet the nation's needs. As the USDA didn't have any in-house experts, the team contacted Richard Harwood, the director of Rodale's Research Center, and our old friend Eliot Coleman, then the director of the International Federation of Organic Agricultural Movements.[3]

The researchers visited many organic farms, both in the United States and abroad, and found that they were seldom as profitable as their industrial counterparts. But they were in better condition: They had healthier soil, less soil erosion, and less water pollution. The team wondered, in writing, whether the profitability data would look any different if the farmers had to cover the costs associated with the declining soil health and erosion and water pollution their methods caused, rather than having taxpayers do it. But actually answering that question was beyond the scope of the study.

Using "a computerized linear programming model" (cutting-edge tech in 1980), analysts tried to calculate the impact if all American farms converted to organic methods. Their conclusion: "Crop production would meet domestic needs, but potential export levels would not be met. Decreased crop production would result in higher farm grain prices and higher farm income in all regions. Total cost per unit of production would be higher, and consumer food prices would be significantly higher." In short, a politically mixed bag—great to have higher farm incomes, not so great for food prices to rise. Good to have plenty for domestic needs, a problem not to have exports.[4]

Acting on one of the report's many recommendations, Secretary Bergland created a position for an organic farming coordinator who was charged with evaluating USDA policies to make sure they were fair to organic farms. But many USDA staffers considered that an attack on chemical-intensive agriculture. Six months after the report came out, Ronald Reagan succeeded Jimmy Carter as president, and Reagan selected John Block as his Secretary of Agriculture. Block ordered all remaining copies of the report destroyed, fired the organics coordinator, and told those working on the report's recommendations to stop.[5]

Outside the USDA, in sharp contrast, interest in organics was strong and growing. Colleges and universities began creating courses in, and building programs about, alternative agriculture, usually calling them "sustainable" rather than "organic." Environmental and food safety advocates embraced organics as of a piece with their agendas. And consumer demand for organic foods was stratospheric: Sales rose sevenfold in just under a decade, from $174 million in 1980 to $1.25 billion in 1989. But customers who wanted organic options were also frustrated, as they had no way to know if items truly were as labeled. Certification requirements varied from state to state; fraud was rampant; and the array of labels promising an item was "ecologically grown" or "natural" or "wild" or "residue-free" had no legal meaning. Buyers wanted to know for sure what they were getting.

In 1990, to appease consumers, Congress pushed through the Organic Foods Production Act (OFPA), despite strong opposition from the USDA. The OFPA established a legal definition of *organic* and banned all other commercial uses of the word. Rather than being an unalloyed victory for organics advocates, however, the OFPA polarized them. One side was eager for the imprimatur a USDA seal conferred and the markets it would bolster. The other side, comprised of more "purist" farmers and environmental advocates, worried the USDA's definitions would end up being watered down.[6]

Eliot Coleman, decidedly a member of the latter camp, groused, "Now that the food-buying public has become enthusiastic about organically grown foods, the food industry wants to take over." He saw the USDA's definition as a way to "meet the marketing needs of organizations that have no connection to the agricultural integrity organic once represented." Coleman was right about the definition meeting a marketing need, as its intent was to let shoppers know that an item was produced organically, which isn't inherently a bad thing. Unfortunately, he was also right that the USDA certification would favor the "food industry." After more than a decade of internal disagreements and politicking, the USDA released standards that were weaker than most advocates wanted and a certification process that was expensive and time-consuming to pursue.[7]

A robust organic-farming sector could have been the leading edge of a transition away from a monopolistic food system. Instead, the certification process and weak rules have led to its absorption, for the most part, into the mainstream system. Not only have CAFOs been allowed to be certified organic, the central role soil plays in organic agriculture is no longer acknowledged within the regulations. As hydroponic produce became popular, growers touted it as organic. After years of urging the growers not to do so, the National Organic Standards Board reversed course and voted to allow hydroponics to be certified as organic—even though hydroponic methods are not soil based. (To be clear, I'm all for incorporating hydroponics into our farm system, but calling a regimen without soil "organic" is like calling a bird a mammal. It's simply not accurate.)

Overall, few US farms are certified organic; only about 1 percent of large farms have become certified. In contrast, among young farmers working at a small scale, certification is more common: Seventeen percent of farmers under the age of thirty-five reported that their farms are USDA certified, and many others report that they're "certified naturally," "certified humane," or something else to indicate their practices are gentler. Some forgo official labels for cost reasons. Vera Fabian and Gordon Jenkins, who run Ten Mothers Farm, explained that at just an acre, their farm is too small for certification to make financial sense. They describe Ten Mothers as "uncertified organic."[8]

Before leaving office, in 1981, Secretary of Agriculture Bergland issued a report that was really a plea, a call for the nation to develop a food and farm policy that would guarantee nutritious food for all. In the portentously titled "A Time to Choose," he argued that America's existing agriculture practices were ruinous, that the country should change its trajectory and ensure reasonable incomes for farmers, fairness across the sector, expanded conservation measures, broader access, and a way of life for farm folks consistent with "other basic goals of our society"—including a "belief in the dignity and worth of all, rewarding the striving for excellence as long as it is not at the expense of others' dignity and survival, promoting access to opportunity, and equity in the distribution of resources, rewards, and burdens, [and] cooperation and shared responsibility." Much as the incoming administration rebuffed the USDA's recommendations about organics, it also ignored this entreaty.[9]

Two decades after Bergland's "A Time to Choose," Secretary Dan Glickman charged yet another USDA commission to develop yet another policy for small farms and issued yet another clarion call, this one entitled "A Time to Act." Acknowledging the department's longstanding failure to adequately support small farms, its authors wrote, "Government policies and practices have discriminated against small farm operators and poor farmers. In some cases, such as commodity program policies, this discrimination was explicit. In other cases, the bias was less intentional and reflected simple ignorance of the specific needs of small farms." Similar to Bergland's report, Glickman's 1998 report said, "Federal farm programs have historically benefited large farms the most. Tax policies give large farms greater incentive for capital purchases to expand their operations. Large farms that depend on hired farmworkers receive exemptions from Federal labor laws allowing them the advantage of low-wage labor costs." The authors described American agriculture as at a turning point: Status quo farming would be extremely harmful, and strong USDA leadership would be necessary for other methods to thrive. Not mincing words, the authors warned, "If we do not act now, we will no longer have a choice about the kind of agriculture we desire as a Nation."[10]

They did not act then.

ECONOMICS: THE TRUE COST OF FOOD

GIVEN HOW CLEAR AND urgent the dangers of climate change have become, I hoped Secretary of Agriculture Tom Vilsack would begin to steer the USDA toward the more environmentally healthy practices the department sidestepped first forty, then twenty years ago. But a year into his term, he continues to support the status quo. Vilsack has been openly dismissive of the Farm to Fork program (F2F), part of Europe's Green Deal. Through F2F, the European Union hopes to move toward a sustainable food and farming system that would also help mitigate climate disruption. Central to those efforts are to reduce reliance on fertilizer and pesticides for produce and the use of antibiotics for livestock. F2F also aims to cut food waste in half and to encourage people to eat more produce and less meat and processed food. To achieve these goals, F2F contains benchmarks by which specific reductions need to occur over the next several decades.

Instead of partnering with Europe in these measures, Vilsack is gathering a coalition of nations united in *not* pursuing the European path; his vision for the United States' agricultural sector is to let the market drive behavior, rely on fossil fuels, and focus on commodities. One of Vilsack's criticisms of F2F is that it would lead to higher food prices and lower production. According to one USDA report, prices could rise as much as 89 percent and production fall 11 percent—if, that is, every single nation adopts F2F and not one innovation in food production or distribution occurs between now and 2050. But as many a Millennial grower or food artisan could tell Vilsack, the reason we pay more for organically produced foods, particularly those from small-scale farms, is because their prices aren't artificially suppressed the way those of industrially produced goods are.

Remember the 1979 USDA study I mentioned a few pages back? The one in which the authors wondered what would happen to the price of conventionally grown foods if the cost of "soil erosion and sedimentation, depleted nutrient reserves, water pollution from runoff of fertilizer and pesticides, and possible decline of soil productivity" were paid for by the farmers? The answer is obvious: Their costs would go up. If they wanted to stay in business, they'd need to raise their prices. In 1979, the USDA team didn't calculate how much those costs would have been. But contemporary analysts have started doing just that, and their results reveal what they call the "true cost" or "full cost" of food.

True cost accountants figure out the costs to remediate social and environmental harms caused by producing something and the value of social and environmental benefits generated by producing it. Then, they add the costs to the sticker price and subtract the value of the benefits. Thus: true cost = sticker price + harms - benefits. If the benefits created outweigh the harms caused, the true cost is lower than the sticker price. A bargain. But if the harms outweigh the benefits, the true cost is higher than the sticker price. A rip-off.

In 2014, the professional services group KPMG calculated the true costs associated with companies across eight business sectors, and found that all of them hid some costs. On average, they hid 41 percent of their costs relative to their pretax profits, which means for every dollar of pretax profit they had, we covered $0.41 of what really should have been their expense. Food producers hide by far the most such costs—224 percent. We paid $2.24 of hidden costs on every dollar of pretax profits those companies made. If the companies had been required to pay for those hidden costs instead of pushing them onto others, they'd have been out of business two times over.[11]

I know this sounds super-wonky. The key takeaway is that true cost accounting reveals how much we *really pay*—not just the amount we hand over at the cash register, but also everything we pay in the forms of taxes, disaster-relief distributions, remediation for environmental harms, climate disruptions, lost opportunities for people, healthcare costs, and economic inequality. The United Nations has developed a sophisticated method for calculating these costs, one that's beyond my mathematical skills.

Fortunately, the food futurist Mike Lee created some props to make true cost accounting clearer. One is a cash-register receipt for a box of Acme cereal. The sticker price for his Acme cereal is $5.29. But the receipt for it includes a few other items: Customers also pay $0.63 to cover the "monoculture soil nitrogen loss" due to soil erosion; $0.66 in a "pesticide hazard surcharge" to cover health costs and lost earnings as farmworkers get sick from pesticide exposure; a "GMO corn license fee" of $2.25 because almost all large farms plant GMO corn; and a "corn subsidy refund" of $0.13 because Americans subsidize farmers growing corn. That box of Acme cereal actually cost the consumer $8.96, close to 70 percent above the sticker price—with just four of the hidden costs added back in.[12]

If real food labels contained the kind of information Lee's did, we'd have a better sense of how much we pay for goods. But as we've seen, going all the way back to the Gilded Age, getting labeling legislation passed in the United States is no easy task. If true cost accounting were applied to US agriculture, we'd see that we already pay more than F2F is likely to cost.

ECONOMICS: TRIPLE BOTTOM LINE

MANY MILLENNIAL FARMERS USE a modified version of true cost accounting called "triple bottom line accounting," in which people, planet, and profit are all considered in the bottom line, not simply profit. To get a crash course in how triple bottom line accounting works on a farm, I visited Kelsey Herrington, of Two Farmers Farm, in Scarborough, in southern Maine. Probably thanks to having an undergraduate minor in economics, she understands and speaks about its importance in especially sophisticated terms.

Before establishing Two Farmers Farm in 2011 with her husband, Dominic Pascarelli, the pair attended Clark University, where Kelsey pursued a master's degree in environmental studies and policy. For her final academic project, Kelsey told me that she looked at the barriers to entry to farming for folks like Tyler Sage, people who want to raise meat sustainably and ethically at a small scale. Starting out in meat farming is expensive; livestock require more infrastructure than does produce, and there are fewer resources for learning about animal husbandry. Kelsey found that a lack of financing was by far the biggest barrier to entry for aspirants who weren't from farm families. But even those who did get a toehold faced many other barriers to success—for example, not having family support for the idea, not having a farmer network, not having the chance to learn from good farmers, not having the same political beliefs as other farmers in the region, or not being white. "I was shocked and upset by how many obstacles there were to do this well and make a living for your family," she said, stunned that even meat farmers with lots of business savvy often live on the financial edge.[13]

As Kelsey's interest in food systems developed, she assumed she'd have a career in public policy, though she and Dominic had plenty of conversations about whether it was better to try to change the system "from within or without." One deterrent to working within was the proliferating technology of that moment, which Kelsey considered "overwhelming. Smart phones," she said, "were just coming out when I was in college. I remember talking to my peers about wanting to do something 'real' or 'tangible.' . . . I remember feeling like we were all becoming a little bit more disconnected from each other and from what really mattered." So she and Dominic took a hiatus from their studies to do something "tangible."[14]

Apprenticing on a year-round vegetable farm in upstate New York, they discovered vegetable farms are somewhat less risky than are livestock farms, but, as Kelsey told me, "there are still a lot of barriers to entry, and there are still a lot of issues with willingness to pay among consumers, and education among consumers." Nevertheless, the pair decided to replicate the four-season vegetable model in southern Maine, where Dominic's family lived. His parents had moved to the state as back-to-the-landers in the 1970s and were supportive of the couple's goals. Kelsey and Dominic also tapped into the region's rich ecosystem of services for new farmers—including MOFGA's journeyperson program, Slow Money Maine's "No Small Potatoes" investment fund, a land match through Land for Good, and business guidance from the Carrot Project.

From the outset, Dominic and Kelsey were committed not just to farming, but also to the triple bottom line. Their mission statement is practically the definition of that philosophy: "Kelsey and Dominic established Two Farmers Farm with a vision for a truly sustainable local food economy. At Two Farmers Farm, 'sustainability' means a high quality of life for farmers and agricultural workers, excellent ecosystems stewardship, a dominance of family-scale businesses, and preservation of agricultural landscapes." In accordance with that mission, they farm organically, pay themselves and their employees a living wage, and charge as little as possible for their produce. When they were starting out, in order to keep expenses down they lived in an apartment above Dominic's parents' garage and rented farmland. After several solvent years, they felt ready to scale up. In 2019, when Kelsey received a MOFGA scholarship, the couple decided to put it toward a root harvester, specifically so they could "replace low-skilled jobs with high-skilled jobs [that provide] more pay and benefits."[15]

As the number of small farms like theirs continues to grow and grocery stores offer more and more organic food, the relative ubiquity of local and/or organic produce puts pressure on businesses like Two Farmers Farm to lower their prices. However, doing that would force them to violate their goal of paying themselves and their employees fairly. So Kelsey and Dominic strategically balance their offerings. They sell some items that command a high price because of their unusualness along with those they can price closer to what supermarkets charge. The pair are adamant about keeping items affordable; working for the triple bottom line means working on behalf of the entire community, not just its wealthiest members. Selling at the minimum viable price is, as Kelsey said, "a way to give back to your community." Even so, she says she's found "a lot of issues with willingness to pay among consumers." To counter that, she gently educates folks at farmers' markets about why local, organic items cost more.

What farmers like Kelsey and Dominic do through enacting their values in their business model, the Greenhorns do by literally acting out their values—that is, by performing them. The Greenhorns were organized in New York's Hudson Valley in 2007, around the same time as the National Young Farmers Coalition, and I think of them as the yin to the NYFC's yang. The group was founded to "promote, recruit and support new farmers in America." Cannily, its first project was to make a documentary about young farmers—a higher-tech version of Helen Nearing's PR efforts to wrangle more "recruits." The Greenhorns describe themselves as a "grassroots media collective" and a "bunch of punks" who want to "increase the odds for success and enhance the profile and social lives of America's young farmers." They've produced a lot of content (a movie, a weekly radio series, videos, guidebooks, manifestos, and farmers' almanacs) and maintain a farm-hack library with instructions for building appropriate tools, such as a bicycle-powered fanning mill for cleaning and sorting seeds, a weld-free root washer, and an eco-friendly slug trap. When the Greenhorns moved out of the Hudson Valley, they brought their eight-thousand-volume agricultural library to their new home, Reversing Hall, in Pembroke, Maine.

More than any other contemporary group I've encountered, the Greenhorns embody the anarchic spirit of the San Francisco Diggers, as is especially evident in their Sail Freight projects. In 2015, the Greenhorns arranged for eleven tons of Maine farm goods—including beans, onions, grains, seaweed, beeswax candles, apple syrup, and blueberry preserves—to be transported by a sailing ship to Boston. Four traditional Norse boats carried goods down the Kennebec River and met up with a windjammer and a schooner. On North Haven Island, they loaded the goods from those vessels onto the 131-foot wooden schooner the *Harvey Gamage*, which then sailed to Portland to get the remaining ten tons of cargo. Six days later, ship and goods arrived in Boston Harbor.

Trading a few-hour highway drive for a weeklong sea journey wasn't simply a diesel-free way to move some food. The Freight Sail, accompanied by teach-ins, concerts, picnics, and other events, was a participatory art performance, economic critique, and world-making effort all rolled into one. Like the original Diggers and the San Francisco Diggers, the Greenhorns were reclaiming the commons and protesting unjust farm practices: "The ocean is a commons not just for commerce. The wind is a commons not just for aviation" they declared in *Mainifesta*, the text accompanying this project. Its preface concludes on an even more emphatic, Digger-ish note:

> We are not content to labor where 70% of the agricultural work is performed by those without citizenship. We are not content to operate in a high-volume, low-value commodity extraction economy. We are not content to be silent while our nation negotiates yet more free trade agreements freeing only those at the top of the capitalist slag heap and chaining the rest of us to their terms. This project is our retort!

(In truth, that project was but one of their many retorts.)[16]

A BEAUTIFUL FARM LIFE

EVEN WITH ALL THE obstacles new farmers face with respect to finding good land and building a viable business, the possibility for small-scale agroecological farms to succeed is greater now than it's been for a century—thanks to the Millennial farmers themselves, the strong consumer interest in healthy, local food, and the infrastructure their seventies-era predecessors built. But farming is precarious. Knowing that, Vera Fabian and Gordon Jenkins, the couple at the heart of Ten Mothers Farm, said that much as they admire their hip forebears, they've tried to be "a little more hard-headed . . . and a little more practical." Rather than set out equipped only with lofty ideals and crossed fingers, the couple turned to some of the most successful food growers and advocates of the previous back-to-the-land generation for guidance. Now, as they share their predecessors' ideas and methods with fellow Millennial farmers and their harvest with CSA members, Vera and Gordon add another link to the daisy chain of food-centric utopians that began in the 1840s.

I met Vera at the Young Farmers Conference at the Stone Barn Center in New York in December 2018. Along with her husband and some friends, she had just purchased twenty-five acres of bare land. Come the new year, they'd begin transforming it into homes, gardens, and Ten Mothers Farm. Although visibly overwhelmed, shaking her head incredulously at the prospect of all the work ahead, she also exuded astonishment and joy about taking this next step toward their collective dream.

A year later, I flew from snowy Maine to a barely balmier North Carolina to spend a day on the farm. When my taxi pulled up to a field of waist-high weeds and dun-colored grasses, the driver asked me to double-check the address. It was the one I'd written down. We spotted the beginning of a dirt road or driveway, one car wide, and I asked him to take it. After about a tenth of a mile, he stopped. The road appeared to dead-end in the tree line. Although I was willing to get out and continue looking for the farm on foot, he was adamant that he wouldn't leave me half an hour from the city in an empty field. So we compromised: He agreed to keep driving until the road was unpassable; I agreed to go back to Durham if the road petered out before we found the farm. As he inched us forward, the road curved again, opening to a vista that included a row of hoop houses. The driver seemed as happy as I was.

Like many others of the new generation of farmers, neither Vera nor Gordon grew up hoping to farm. Vera's strongest "farm memories" from childhood are all from her favorite picture books. But she did grow up loving food, so much so that she learned to cook at age four and is now, in her own words, "a little food obsessed." In college, she learned about food and food systems from many vantage points, taking classes from different departments to cobble together her own food studies curriculum because the university didn't have one. As she did so, she also began to be aware of environmental issues and helped start an environmental education program for elementary school students that used foods to foster connections.

But until her junior year, when she spent a semester in Africa, she'd never grown food. In Mali, Vera worked in a women's market garden. The women grew food for their families and sold their surplus, using the proceeds to send their children to school. Vera loved the way the garden solved the immediate problems the women faced and says she realized that she "could make the world more the way I want it to be through growing food and through education . . . This women's garden was very much about women teaching each other." When she returned to the States, she heard about the Edible Schoolyard that Alice Waters spearheaded at a middle school in Berkeley. She applied for a position there because she wanted "to learn from people who are good at what they do." In 2007, straight out of college, Vera landed her dream job.

Gordon Jenkins, who grew up in Berkeley, didn't have the same long-term passion about food that Vera had, though he does remember stumbling on the Japanese predecessor to *Iron Chef* sometime during high school and organizing *Iron Chef* competitions with his friends—mostly, he told me, because he found them hilarious. Whatever interest in food those competitions may have triggered was quickened appreciably when he reached college. At Yale, he and another freshman started a tea club; to their surprise, it took off. Wanting to be a good host, Gordon thought he ought to know something about tea. "And that," he said drolly, "was the beginning of the rabbit hole, and I fell down it. And got *really* into food." During sophomore year, a friend brought him to Yale Farm, a one-acre urban farm the university runs as part of the Yale Sustainable Food Program, which Alice Waters had designed a few years earlier. He did a summer apprenticeship on the farm and "fell in love with farming."[17]

"I enjoyed the work; I felt alive doing it," he said. "I read Michael Pollan and got really into Slow Food. Everything those people were talking about I was doing every day. And it connected. I felt like I should be doing this." The Yale Farm was run organically, which Gordon knew was good for people and planet: "It was the kind of food system that, in theory, would be more fair for everyday people, too. So it had a social-justice aspect even though that wasn't an explicit intention of that farm."

After graduation, Gordon returned home to Berkeley and took a job as Alice Waters's assistant. Fortuitously, that required spending some time at the Edible Schoolyard at the Martin Luther King Jr. Middle School, where Vera was working. "Our first conversation was about how we dreamt about farming," she recalled, "but neither of us had the courage to do it straight out of college. The world had told us we needed to do something 'more important' than farming. You know, we were 'too smart' for farming, we had college degrees. It took us years to work up the courage to actually do it." Working in a series of educational gardens, first in Berkeley and later in New York, Vera was able to somewhat sate her desire to farm, though she wished, as she said, that she was nurturing "a deep, long-lasting connection to a piece of land." Gordon's jobs, though food-adjacent, were all indoors.

In 2008, under the auspices of Slow Food USA, Alice Waters set out to produce "Slow Food Nation," a weekend-long gathering to celebrate local, slow, organic food. A quarter-acre of the plaza in front of San Francisco's City Hall had been dug up the previous spring and planted as a Victory Garden. Rimming the plaza was a sixty-vendor farmers' market. Lectures and panel discussions highlighted a broad array of topics, among them local and sustainable food practices, edible education, the global food and energy crises, inhumane working conditions for farm laborers, and heirloom foods. A suite of "taste pavilions" featured foods prepared by well-known chefs and assorted artisanal fare. The goal of the convening, which Michael Pollan called Slow Food USA's "coming-out party," was to inspire a revolution to improve America's food culture, making it not only healthier but also more just. Between fifty thousand and sixty thousand people showed up.[18]

Gordon was among those who pulled together the pieces. He organized the farmers' market as well as an "eat-in" at Dolores Park. The eat-in blended a DIY zeal much like that of the Diggers with a white-tablecloth sensibility. A sinewy table long enough to seat three hundred, clad in white linen and beautifully laid out, stretched through the park. Nearby was a makeshift stage: The low platform, supported by bales of hay and framed by a bright red archway, recalled the Free Frame of Reference folks stepped through at Digger Feeds four decades earlier. The eat-in was a potluck, with dishes that incorporated lots of leftovers from the farmers' market and produce from the Victory Garden.

That eat-in inspired others. The following year, more than three hundred eat-ins took place on Labor Day to launch Slow

Food's "Time for Lunch" campaign. But as Gordon's responsibilities at Slow Food USA grew, chances to pursue such playful activism waned. When the organization was going through significant change, and Gordon felt inundated with meetings and extra work, he'd reread a quote from the theologian Howard Washington Thurman that he kept over his desk: "Don't ask yourself what the world needs. Ask yourself what makes you come alive, and go do that, because what the world needs is people who have come alive."

Though Gordon and Vera were both pretty sure farming would make them most alive, they were frustrated by their own mix of yearnings and paralysis, and they made a deal not to talk about farming for a year. On January 1, 2012, having kept their promise, they discovered they'd both read a ton about farms the previous year. That day, they began their plan. Their route forward began by looking backward—to three iconic figures from the last back-to-the-land moment: Alice Waters, Bob Cannard, and Eliot Coleman. They apprenticed first with Bob Cannard, of Green String Farm "because," as Vera said, "he had been the farmer for Chez Panisse. We knew his vegetables were the most delicious," and then with Eliot Coleman "because Eliot's book is the first book we'd ever read, years before."

The information and methods Bob Cannard and Eliot Coleman covered were similar, but for Vera and Gordon, the cultural aspect of being at Four Season Farm was important in and of itself. As lifelong urbanites, they weren't entirely sure they'd be comfortable living somewhere remote, with few friends or neighbors. In Harborside, Vera said, they found themselves "in a quiet place and relying on just a handful of people for company and stimulation . . . There was a kind of withdrawal period to get used to it, but we came to really like it. That community is still sort of a model for how we want to live in North Carolina."

Wanting to live close to both friends and family, they looked for land in either the Sonoma area or in North Carolina. They

soon settled on the latter, in part because it was much more affordable. But, as Vera explained, they were caught off guard by
"how hard the climate is for growing produce there." To gain
some local knowledge, they apprenticed at an organic farm in
Cedar Grove, about twenty miles northwest of Durham, before
going out on their own. Partway through that apprenticeship,
they got married. Vera laughed when she told me, "The whole
wedding registry was for things like an orchard for our future
farm and hand tools for our future farm."

They established Ten Mothers Farm on rented land while
they looked, with friends, for a parcel to buy for what Vera calls
their "forever farm." In 2018, their third year renting, the landowner decided to sell. They had to leave at the end of the year,
and they had seventy CSA members whom they didn't want to
disappoint. As Vera talked about this, I flashed back to the day
I'd met her and the mix of joy and panic she'd exuded about having bought land. They'd found it just in the nick of time.

When the taxi and I reached Ten Mothers Farm, Vera was
already picking carrots for that week's CSA share box. With her
were Luke Howerter, a friend who'd been working on the farm
all season, and Justin Nye, who volunteered at Ten Mothers on
Mondays. Justin's job-job is running a restaurant's front of house,
but he spends his day off working at Ten Mothers because, as he
put it, "a bad day on a farm is better than a good day in a restaurant." Gordon was in the farm's cathedral-style high hoop house,
washing carrots from the first box they'd harvested. Digging carrots is something I've had plenty of practice doing, so I joined
them pulling and bundling carrots into one-pound bunches.

Vera and Gordon (and I) set out to cross off a lot of smaller
items on the to-do list while Justin and Luke began working on
one big one—building a second high hoop house. Ten Mothers
needed the additional space so soon because its CSA clientele
had more than doubled. While Justin and Luke got started on
that, we prepared daikon radishes for the CSA share boxes and

uncovered the covered crops in the fields and low hoops to give them some light and air. We spread compost on the new pollinator bed. Amended it. Edged a second pollinator bed. Laid down kraft paper and covered it with compost. And more.

Justin and Luke had all the hoops built by lunchtime, when they discovered a problem: The screws for the collar ties were the wrong length. By the end of day, however, that was resolved and the frame was up, braced, and collared. The hip rails and baseboards would get installed the next day. Over lunch with the crew and Kelly, one of the other "land mates," the conversation turned not to freedom, as it had on several other farms I'd visited, but to beauty. When I'd first met Vera, she said what motivated her to farm were two things: "a desire to prevent us from making our world uninhabitable for environmental reasons—there's that practical sense of urgency around climate change and resources. And then there's the more human part of it, and that's this desire to have the world be more beautiful." She was surprised when I told her I hadn't heard anything about beauty from other farmers, not even from flower farmers or folks with fields so lovely they seemed designed for Instagram.

The more we talked, the more I came to understand that when Vera uses the word *beautiful*, she means something far more comprehensive than "very pretty." "Things that are alive are beautiful," she said, "as are things that generate aliveness or a sensation of aliveness." Tasks on the farm—"even the ugly and hard parts—to me, are beautiful," she added. Gordon suggested she finds these beautiful because she knows what they contribute, where they fit into the whole of what they're doing. I think he's right, for Vera went on to describe her vision of a "more beautiful world" as one with "more tight-knit communities, relying on land and the environment around them, and a smaller group of people they're close to."

Her beautiful life is much like the free one so many other utopians described. But it adds, as the Ardenites' version did and Alice Waters's version does, a regard for aesthetics and a conviction that beauty isn't superfluous, that it's integral to a just, healthy, connected life.

Gordon and Vera are excellent farmer-entrepreneurs carving out what I hope will be a very beautiful life. But it isn't as rosy as the ones in Vera's childhood picture books. This one includes grasshopper infestations, failed crops, cultivars that don't live up to expectations, increasingly unpredictable weather, and significant economic uncertainty. As Vera and I spread compost over one of the new pollinator hedgerow beds, we talked about the financial challenges she and Gordon and other farmers face. She's frustrated that farmers must struggle to make a living and that many people who want to farm in an organic or regenerative way can't—that student loan debt or the price of land or the lack of a large enough customer base compels them to give up their farming aspirations. She's seen plenty of her peers return to the city and to urban jobs because they can't afford to farm, their trajectory a perverse inversion of the one that's been gutting rural America since the 1840s. One of her dreams, she says, is that "by the time our children are grown, farming will be seen very differently."

I hope her dream is realized, especially because the lifestyle and relationship to the land Vera, Gordon, and their land mates are striving to create for themselves and future children is precisely the sort many believe is the best for the planet. In *This Land Is Our Land: The Struggle for a New Commonwealth*, Jedidiah Purdy turned his attention to the importance of land and to how we occupy it. Like the original Diggers, like Sir Thomas More, like Henry David Thoreau and Bronson Alcott and Charles Fourier, like Henry George and Frank Stephens, like Scott and Helen Nearing, like the San Francisco Diggers and so many other hip folks of their era, Purdy has come to regard private landownership as a major source of America's ongoing inequality: "I come back to the land and the thought that it holds people both together and apart," he writes, and to "the idea that it belongs originally and essentially to everyone, that it is a commonwealth."[19]

Taking seriously the notion that land is a source of common health and wealth requires a very different economic model than the one most Americans embrace, one more consistent with true cost accounting than with a conventional profit-and-loss statement. In a commonwealth, Purdy observes, "no one gets their living by degrading someone else, nor by degrading the health of the land or the larger living world. In such a community, the flourishing of everyone and everything would sustain the flourishing of each person. This would be a way of living in deep reciprocity as well as deep equality." This is the vision animating all of the back-to-the-land movements chronicled here, but I can't help thinking it especially resonates with Caitlin Frame's description of "real organics" and one of Vera's definitions of a beautiful life: "being in a place that's a beautiful place and feeling that sense of support from friends and family and neighbors and feeling like everybody has their place and their work to do and is needed."[20]

From Thoreau and Emerson and Thomas Carlyle and the Arts and Crafts movement to Frank Stephens and the Ardenites, to Scott and Helen Nearing, to Nancy Berkowitz and Eliot Coleman and Alice Waters, to Eric Harvey and Vera Fabian and Gordon Jenkins, each generation of back-to-the-land utopians has been personally connected to members of an earlier one, has learned from, admired, and often loved them. These idealists are but a few of the many who have been intimately and integrally connected to utopians who came before or after them, who inspired or affirmed them, who believed that land is a common good, that farming should be a healthful enterprise, that food should nourish bodies and spirits. By enacting their beliefs, by dreaming into being a more capacious world, they changed the existing one for all of us. They've expanded our vision of community to include animals and water and soil as well as other people. They've reminded us that eating is an ethical and potentially delicious, nutritious enterprise. They've worked hard to get adulterated foods out of our cupboards and refrigerators. And they've reinforced our understanding that freedom is a collective attribute, that no one can be truly free if others aren't also free.

ACKNOWLEDGMENTS

JUST TWO MONTHS AFTER I started working on this book, I was diagnosed with breast cancer. As I learned more than I'd ever wanted to know about increased cancer risks associated with various foods, herbicides, and pesticides, the amazing work of contemporary young farmers began to feel especially personal and urgent. Thank you to all the growers, activists, artisans, land-access and farm-justice workers, seed farmers, and others who do the work to create local, resilient, healthy food systems. I am beholden to you, period.

Huge thanks to the farmers and food artisans who took time out of their packed schedules to talk to me about what they do and why: Kate Del Vecchio and Rich Lee, of Tender Soles Farm; Leah Perriman, of Soul Fire Farm; Allison Lakin and Neil Foley, of East Forty Farm; Andy Smith and Caitlin Frame, of The Milkhouse; Diane Einseidler and Eric Harvey, of Harlem Valley Homestead; Vera Fabian and Gordon Jenkins, of Ten Mothers Farm; Elizabeth Siegel, of Heritage Home Farm; Jacinda Martinez, of Grounded Local; Kelby and Pamela Young, of Olde Haven Farm; Kelsey Herrington, of Two Farmers Farm; Laura and Craig Martel, of Greener Days Farm; Ben Rooney, of Wild Folk Farm; Sam Mudge, of Grange Corner Farm; Tyler Sage, of Sage Farm; Joel Alex, of Blue Ox Malthouse; Jenn Legnini, of Turtle Rock Farm; Justin Morace and Nell Finnigan, of Ironwood Organic Farm; and Fernando Orozco, of Roof Crop Farm.

Thanks also to their allies who told me about efforts to transform our food system into something saner and safer, including: Sara Trunzo, from Maine Farmland Trust; Tristan Noyes, from the Maine Grain Alliance; John Piotti, of American Farmland Trust; Patrick Connors and Ben Holmes, of the Farm School; and Doug Fox, from Unity College.

Deep, deep gratitude to the folks who reminisced with and for me about their back-to-the-land experiences in the 1970s: Warren and Nancy Berkowitz, Chris and John Bly, Eliot Coleman, Sherm Hoyt, Les Hyde, US Representative Chellie Pingree, Craig Wilgus, Jay Davis, Levi Walton, and Bonnie Rukin.

Thanks to Jeff Creamer, at the Walden Woods Project, for his help navigating the archive. Shout-outs to Charlotte Sheedy, Susan Bates, and Lewis Robinson for their early support and guidance. Big props to Jeffrey Kittay and Mark Munger for kind and useful feedback on early drafts. *Grazij* to my Breadloaf crew in Sicily, especially Ellen Clegg.

And how to thank the Godine team, especially Joshua Bodwell and Celia Johnson . . . you're the best allies an author could dream into being. You helped transform a free-range tale of every-cool-or-terrifying-thing-I-know-about-food-and-utopia-and-capitalism-and-farming into this true story.

And to Rob, my first reader, best friend, and favorite foodie—thank you, baby.

NOTES

INTRODUCTION

1 Harvey, N., "Why Are Young, Educated Americans Going Back to the Farm?"
2 Conway, *Get Back, Stay Back*, 20.
3 Alcott, L., *Transcendental Wild Oats*, 55–56.
4 Vespucci, "First Voyage."
5 More, *Utopia*, 63.
6 Purdy, *A Tolerable Anarchy*, 13.
7 Ibid., 124, 125.
8 Thoreau, *Walden*, 336.
9 Berry, *The Unsettling of America*, 22.

CHAPTER ONE: MILLENNIALS MOVE TO THE LAND

1 Siegel, Elizabeth, all quotes from interview with the author, April 17, 2017.
2 Algiere, Jack, talk at Stone Barns Center Young Farmers Conference, December 5, 2018.
3 Frisch, "To Free Ourselves, We Must Feed Ourselves"; Penniman, Leah, "A New Generation"; Penniman and Snipstal, "Regeneration," 64.
4 Lozano, "24 LGBTQ+ Farms."
5 Smith, Andy, quotes from interview with the author, May 25, 2017, unless otherwise noted.
6 Fabian, Vera, quotes from phone interview with the author, December 11, 2018, unless otherwise noted.
7 Van Gelder, Sarah, "Slow Food."
8 USDA, 2012 Census of Agriculture. That trend continued. In the 2017 Agricultural Census, the number of farmers under twenty-five, between twenty-five and thirty-four, and between thirty-five and forty-four were all higher than in 2012; National Young Farmers Coalition, "Building a Future with Farmers II," 9.
9 Martel, Craig, interview with the author, April 13, 2017.
10 White, Nora E., "Farming in the time of pandemic."
11 Maine Farm and Seafood Products Directory.
12 Trunzo, Sara, interview with the author, September 22, 2015.
13 National Young Farmers Coalition, "Building a Future with Farmers II," 9.
14 Frame, Caitlin, quotes from interview with the author, May 25, 2017, unless otherwise noted.
15 Nosowitz, Dan, "The Real Organic Project."
16 A large CAFO is a feeding operation with more than a thousand "animal units" (that's the real term). According to the USDA, "[A]n animal unit is defined as an animal equivalent of 1000 pounds live weight and equates to 1000 head of beef cattle, 700 dairy cows, 2500 swine weighing more than 55 lbs., 125 thousand broiler chickens, or 82 thousand laying hens or pullets."
17 Richmond, Todd, "Ag secretary: No guarantee."
18 Frame, Caitlin, "Real Organic Project Symposium"; Frame, "Know Your Farmer."
19 June 5, 2021, Instagram post.
20 James, et al. "Do minority and poor neighborhoods have higher access . . .?"

21 Penniman, Leah, *Farming While Black*, 224.
22 Heffernan, Ashley, "Farm store brings fresh produce back."
23 Jenkins, Germaine, "Nom, Nom, Nom."
24 Holmes, Ben, interview with the author, June 7, 2017.
25 Frisch, "Taking on Food Justice with Soul Fire Farm's Leah Penniman."
26 Frisch, "To Free Ourselves, We Must Feed Ourselves."
27 Ibid.
28 Harlem Valley Homestead website.
29 Harvey, Eric, interview with the author, December 5, 2018.
30 Legnini, Jenn, all quotes and references from interview with the author, December 7, 2017.
31 Rooney, Ben, interview with the author, March 20, 2017; Mudge, Sam, interview with the author, March 29, 2016.
32 Heart of Maine website, wayback machine, June 14, 2007.
33 Stir the Pots blog, April 1, 2020; Wu, Tim, "That Flour You Bought Could Be the Future of the U.S. Economy."
34 Noyes, Tristan, all quotes from interview with the author, May 23, 2017.
35 Alex, Joel, all quotes from interview with the author, April 17, 2019.
36 Evans, Pat, "All 50 States Ranked for Beer."
37 Goad, Meredith, "Grain Partnership."

CHAPTER TWO: HENRY DAVID THOREAU'S SEARCH FOR FREEDOM

1 Carlyle, Correspondence, 308–309.
2 Quoted in Scharnhorst, "Five Uncollected Contemporary Reviews of 'Walden,'" 3; Grant, "*Walden* Wonder."
3 Brand, *Last Whole Earth Catalog*, 37.
4 Quoted in Walls, *Henry David Thoreau*, 76.
5 Brownson, *New Views*, 36, 39, 46, 49.
6 Ibid., 55, 97.
7 Thoreau, "Commencement Address," 9.
8 Quoted in John Matteson, *Eden's Outcasts*, 71.
9 Thoreau, *Walden*, 254, 274.
10 Emerson, Sermon CIV, 84.
11 Ginsberg, "Footnote to 'Howl.'"
12 Thoreau, *Walden*, 249.
13 Quoted in Schreiner, *The Concord Quartet*, 70–71.
14 Hawthorne, journal entry, September 1, 1842.
15 Quoted in Walls, *Thoreau*, 189.
16 Thoreau, *Walden*, 270 (There's some quibbling about who lent Thoreau the ax, as several of his friends have taken credit it, but Alcott is the likeliest candidate); ibid., 273.
17 Ibid., 304.
18 Quoted in Helen Nearing, "Thoreau, Judged in His Own Time," 2.
19 Bidwell, "The Agricultural Revolution in New England," 683.
20 Ibid., 688.
21 Gross, "Culture and Cultivation: Agriculture and Society in Thoreau's Concord," both 51.
22 Ibid., 51, 52.
23 Ibid., 52.

24 Thoreau, *Walden*, 265; Saloutos, "The Agricultural Problem," 158.
25 Quoted in Bidwell, 700.
26 Thoreau, *Walden*, 286.
27 Marx, *Capital*, Vol. I, 637.
28 Thoreau, *Journal III*, March 2, 1852.
29 Thoreau, *Journal I*, July 14, 1845; Thoreau, *Walden*, 247.
30 Thoreau, *Walden*, 261.
31 Ibid., 264. This phrasing wryly echoes Adam Smith's definition of *value*: "The value of any commodity, therefore, to the person who possesses it, and who means not to use it or consume it himself, but to exchange it for other commodities, is equal to the quantity of labor which it enables him to purchase or command" (in *The Wealth of Nations*); Thoreau, *Walden*, 267. (If only glow-shoes were a nineteenth-century precursor to blinking sneakers or eighties-era rave-wear. Alas, they're merely galoshes.)
32 Thoreau, *Walden*, 267 and 266.
33 Ibid., 360; Thoreau, *Journal V*, August 23, 1853.
34 Thoreau, *Walden*, 352.
35 Ibid., 350, 354.
36 Ibid., 354, 355.
37 Ibid., 353, 354.
38 Ibid., 385, 387, 387.
39 Ibid., 388.
40 Ibid., 330.
41 Ibid., 332.
42 Ibid., 337, 331; Maynard, *Walden Pond, A History*, 73; and Walls, *Henry David Thoreau*, 194.
43 Lemire, *Black Walden*, 10.
44 Walls, *Henry David Thoreau*, 215; Conrad, "Realizing Resistance," 167.
45 Quoted in Dedmond, "The Root of the Hydra's Head," 3.
46 Thoreau, "Slavery in Massachusetts."
47 Thoreau, *Journal XII*, October 19, 1859, and October 22, 1859.
48 Thoreau, *Journal VIII*, November 5, 1855.

CHAPTER THREE: SECULAR UTOPIAS OF THE 1840s

1 Guarneri, "Reconstructing," 463.
2 Anderson, J., "Fruitlands," 668.
3 Alcott, L., *Transcendental Wild Oats*, 69.
4 Shepard, *Journals*, 69; Alcott, A. B., *Table-Talk*, 22.
5 Matteson, *Eden's Outcasts*, 102.
6 Sanborn, ed., *Bronson Alcott and Alcott House*, 17–18.
7 Adams and Hutter, *Mad Forties*, 24.
8 Alcott and Lane, *The Dial*, 136.
9 Rodabaugh, "A Nation of Sots," 29.
10 Quoted in Wallach, *How America Eats*, 150–151; "American Vegetarian Convention: A Detailed Report."
11 Graham, *A Treatise on Bread and Bread-Making*, 19–20.
12 Ibid., 44, 45, 47.
13 Sanborn and Harris, *Bronson Alcott*, 379.
14 Sears and L. Alcott, *Bronson Alcott's Fruitlands*, 44; Matteson, *Eden's Outcasts*, 110, 112.

15 Alcott, L., *Transcendental Wild Oats*, 28.
16 Ibid., 34–35, 35.
17 Ibid., 40–41; Sears and L. Alcott, *Bronson Alcott's Fruitlands*, 115.
18 Alcott, L., *Transcendental Wild Oats*, 41.
19 Ibid., 36.
20 Sears and L. Alcott, *Bronson Alcott's Fruitlands*, 42.
21 Quoted in Matteson, *Eden's Outcasts*, 114; Emerson, Journal, July 8, 1843; Carlson, "The Inner Life of Fruitlands," 100.
22 Guarneri, "The Americanization of Utopia," 78; quoted in Delano, 34.
23 Quoted in Zonderman, 196.
24 Gregory, "Marx's and Engels' Knowledge," 181, 157.
25 Guarneri, "The Americanization of Utopia," 79; Guarneri, "Reconstructing the Antebellum," 467; Guarneri, "The Americanization of Utopia," 77.
26 Zonderman, "George Ripley's Unpublished Lecture," 189.
27 Fourier, *The Theory of the Four Movements*, 264.
28 Levi, *Food in Utopia*, 158.
29 Quoted in Levi, *Food in Utopia*, 152.
30 Brown, J., "Alimentary Discourse," 86, see also footnote 12; Levi, *Food in Utopia*, 168.
31 Fourier, *The Four Movements*, 87; Levi, *Food in Utopia*, 163.
32 Matthews, "An Early Brook Farm Letter," 229.
33 Quoted in Levi, *Food in Utopia*, 171–72, 171; Delano, 52.
34 Reed, *Letters from Brook Farm*, 41.
35 Sargent, "Social and Political . . .," 44, 58.
36 "The Question of Slavery," *The Harbinger*, 30.
37 Petrulionis, "Swelling That Great Tide of Humanity," 399.
38 Ibid., 401.
39 Levi, *Food in Utopia*, 172.

CHAPTER FOUR: GOING BACK TO THE LAND DURING AMERICA'S FIRST GILDED AGE

1 Both signs are allusions to Shakespeare—the welcome comes from *King Lear* and the farewell from *Julius Caesar*.
2 Phillips, *Wealth and Democracy*, 210, xvi.
3 Epstein, "Have American Wages Permitted an American Standard of Living?" 189; US Bureau of Labor, "Retail Price of Food, 1890–1906," 191.
4 George, *Progress and Poverty*, 7.
5 Ibid., 9, 10.
6 Edwards, "Arden," 22.
7 Ibid., 65.
8 Quoted in Edwards, "Arden," 27; Morris, preface to "The Nature of Gothic," i.
9 Thomas, *William L. Price*, 70.
10 Stephens, "The Arden Enclave," 72; Taylor, 316.
11 Wiencek, "A Delaware Delight," 9.
12 Nearing, S., *Frank Stephens' Songs*, unpaginated.
13 Thomas, *William L. Price*, 42.
14 Stickley, *More Craftsmen Homes*, 7; Thomas, *William L. Price*, 95.
15 Stickley, *More Craftsmen Homes*, 9.
16 Taylor, "Utopia by Taxation," 318, 317; Newhall, "Arden—A Colony of Pleasure and Profit," 75.

17 Magliari, "Free State Slavery," 157.
18 Burnett, "State of the State."
19 Hurt, *American Agriculture*, 184.
20 Sherman, "Special Field Order Number 15."
21 Blum, *The Poison Squad*, 13.
22 Mason, "Food Adulterations," 552; Blum, *The Poison Squad*, 87.
23 Sinclair, *The Jungle*, 56.
24 Wiencek, "A Delaware Delight," 10.
25 Walker, "Delaware's Odd, Beautiful . . ."
26 "Radical professor recalls Arden life of 50 years ago," 9.

CHAPTER FIVE: SCOTT AND HELEN NEARING LIVE THE GOOD LIFE

1   Gray, "Radical Teaching," 269.
2   Young, Kelby, and Pam Young, interview with the author, April 5, 2017. All quotes from this interview.
3   Nearing, S., *The Making of a Radical*, 36; Buhs, "Scott Nearing as a Fortean."
4   Nearing, S., *The Making of a Radical*, 38–40. The other two were the Child Labor Committee and a luncheon club organized by the Philadelphia Ethical Society.
5   Nearing, S., and Nellie Seeds Nearing, *Woman and Social Progress*, xi.
6   Nearing, S., *Reducing the Cost of Living*, 91; Phillips, *Wealth and Democracy*, 38. The economic gap was not this stark again until the end of the 1990s. You may be wondering how stark the divide is now: At the end of 2020, Jeff Bezos was 1.94 million times wealthier than the median American household.
7   Nearing, S., *Reducing the Cost of Living*, 217–18, 218, 219.
8   Nearing, S., "The Adequacy of American Wages," 111, 118.
9   Killinger, *Helen Knothe Nearing: A Biography*, 13.
10 Killinger, *The Good Life of Helen*, 33; and Killinger, *Helen Knothe Nearing*, 58.
11 Killinger, *The Good Life of Helen*, 34.
12 Ibid., 37; Nearing, H., *Loving and Leaving*, 79, 93.
13 Wallach, *How America Eats*, 102–103.
14 Nearing, S., *The Making of a Radical*, 158, 160.
15 Ibid., 210.
16 Nearing, H., *Our Home Made of Stone*, iii.
17 Miller, *The Quest for Utopia*, 121.
18 Borsodi, *This Ugly Civilization*, 136.
19 Ibid., 272.
20 Nearing, S., "The Relation of Simple Living to Social Radicalism," 4, 6.
21 Nearing, H., and S. Nearing, *The Good Life*, 4.
22 Nearing, H. and S., "Raising Children."
23 Nearing, S., "Scott Nearing on The Good Life"; Nearing, S., *The Making of a Radical*, 215, 216.
24 Both from Johnson, "Nearing Enough."
25 Nearing, S., "Arden Town," unnumbered synopsis page.
26 Ibid., 23.
27 Ibid., 37–38, 39, 40, 40, 40.
28 Ibid., 41, 42.
29 Ibid., both from 138.
30 Ibid., 154, 202, 205.

31 Newhall, "Arden—A Colony of Pleasure and Profit," 75; McKinney, "What the Arden School Can Teach Us," 7.
32 Newhall, "Arden—A Colony of Pleasure and Profit," 76.
33 Nearing, H., and S. Nearing, *The Good Life*, 200.
34 Ibid., 200–201, 201.
35 Nearing, H., untitled public lecture.
36 Nearing, H., *Simple Food for the Good Life*, 225.
37 Ibid., 226, 227.
38 Johnson, "Nearing Enough"; Nearing, H., and S. Nearing, *The Good Life*, 127; and Nearing, S., *The Making of a Radical*, 227.
39 US FDA, "80 Years of the Food, Drug and Cosmetic Act"; Poches, "The American Chamber of Horrors."
40 Nearing, S. *The Making of a Radical*, 241; Thoreau, *Walden*, 360.
41 Nearing, H., interview with Nina Ellis.
42 Berkowitz, N., interview with the author, September 20, 2016. Quotes from this interview unless otherwise noted.
43 Killinger, *The Good Life of Helen*, 95; Nearing, H., and S. Nearing, "Summing Up," 62.
44 Nearing, H., and S. Nearing, "Summing Up," 60.
45 Ibid., 60; Nearing, H., untitled public lecture.
46 Bodwell, "Year-Round Radicals," 65; Coleman, E., interview with the author, April 11, 2017. Quotes from this interview unless otherwise noted.
47 Bright, *Meanwhile, Next Door to the Good Life*, 24.
48 Nearing, H., and S. Nearing, "Summing Up," 60.
49 Bright, *Meanwhile, Next Door to the Good Life*, 289, 290.
50 Ridley, "Back to the Land 2.0."
51 Conway, 140, 142.
52 Ridley, "Back to the Land 2.0."

CHAPTER SIX: FOOD FIGHTS

1 Ziegelman and Coe, *A Square Meal*, 153–154.
2 Culver and Hyde, *American Dreamer*, 115.
3 Ibid., 124.
4 Ibid., 125.
5 Gunderson, "The National School Lunch Program"; Levenstein, *Revolution at the Table*, 119.
6 Ziegelman and Coe, *A Square Meal*, 80–81.
7 Ibid., all on 82.
8 Levine, *School Lunch Politics*, 68, 68, 69.
9 Elzey, "An Elementary School Garden Project," 115; Heuchling, "Children's Gardens in Chicago," 105; and US Bureau of Labor, "City Gardens in Wartime," 644.
10 Halper, "Evolution and Maturity of the American Supermarket," 256.
11 Curtis, "Modern Kitchen Miracles," C2.
12 Hamilton, S., "The Economies and Conveniences of Modern-Day Living," 36, 37.
13 Neuhaus, "The Way to a Man's Heart," 533; Shapiro, *Something from the Oven*, 27.
14 Barber, *The Third Plate*, 50; Petty and Schultz, "African-American Farmers and the USDA," 339.

15 Hurt, *American Agriculture*, 307.

16 Rodale, *Pay Dirt*, 67.

17 Leopold, *Sand County Almanac*, 203, 204.

18 Goldman and Hylton, eds., *The Basic Book of Organically Grown Foods*, 328.

19 Rohman, "Let Us Boycott Lettuce"; "Farm Workers Press Lettuce Boycott."

20 Oakes, K., *Slanted and Enchanted*, 34.

21 Diggers, "Let Me Live in a World Pure."

22 Grogan, *Ringolevio*, 283.

23 Ibid., 284, 286.

24 Coyote, *Sleeping Where I Fall*, 70–71.

25 Wallach, ed., *American Appetites*, 198.

26 Diggers, "Trip Without a Ticket" and "Men Are Out of Touch with the Earth."

27 Forman, "San Francisco style."

28 Weller, "Suddenly That Summer," 77.

29 Patel, "Survival Pending Revolution," 2.

30 Houriet, *Getting Back*, 17; Kauffman, *Hippie Food*, 184.

CHAPTER SEVEN: HIPPIE FOOD

1 Muhlke, "The Hippies Have Won."

2 Quoted in Kamp, *Arugula*, 185.

3 Starkman, "The Radical Necessity of Cooking."

4 Houriet, *Getting Back*, 10, 11.

5 Ibid., 22.

6 Ibid., 22; Houriet, "Life and Death," 100.

7 Houriet, *Getting Back*, 173; Smith, P. "Brotherhood Spirit"; Sokolov, "The Food at the Heart," 60.

8 Quoted in Peter Smith.

9 For a fuller account, see Belasco, *Appetite for Change*.

10 Bloodroot Collective, *The Political Palate*, xi, xii; Bloodroot website, "about."

11 Horton, *Country Commune Cooking*, 81.

12 Atwater, *Natural Foods Cookbook*, 104, 104–105, 105.

13 Kauffman, *Hippie Food*, 61.

14 Hannaford, 2, 3, xix.

15 Kauffman, *Hippie Food*, 80.

16 Hartman, "The Political Palate," 30.

17 Dragonwagon, *The Commune Cookbook*, 68.

18 Hartman, "The Political Palate," 36.

19 Brock, *Alice's Restaurant Cookbook*, 25.

20 O'Sullivan, *American Organic*, 87.

21 Quote in Sokolov, 60.

22 Walton, L., interview with the author, February 25, 2016. All quotes from this interview.

23 Davis, J., interview with the author, September 22, 2015. All quotes from this interview.

24 Wilgus, C., interview with the author, November 11, 2015. All quotes from this interview.

25 Hoyt, S., interview with the author, September 16, 2015. All quotes from this interview.

26 Bly, C. and J., interview with the author, October 15, 2015. All quotes from this interview.

27 Coleman, M., *This Life Is in Your Hands*, 65.
28 Horton, *Country Commune Cookbook*, 14.
29 Darlington, *Grow Your Own*, 1–2, 2.
30 Levenstein, *Fear of Food*, 117.
31 Nader, "Watch That Hamburger," 15; Levenstein, *Fear of Food*, 117.
32 Kauffman, *Hippie Food*, 117; Belasco, *Appetite for Change*, 48 and 49.
33 Dragonwagon, *The Commune Cookbook*, 127–28, 129.
34 Coleman, M., *This Life Is in Your Hands*, 78–79.
35 Belasco, "Food and the Counterculture," 286.
36 "Lucy Horton Interview."
37 Murray's Cheese Blog.
38 Kamp, *United States of Arugula*, 170; Jacobs, *New Pioneers*, 122.
39 Quoted in Kamp, 171; Laura Chenel, two quotes from "40 Years of Love (and Goat Milk)."
40 Cypress Grove Cheese, both quotes from "Our Story."
41 American Cheese Society, "About Us."
42 Carroll, "Guide to Home Cheese Making."
43 New England Cheese Making Supply Company, "Our Story."

CHAPTER EIGHT: TRANSFORMING THE FOOD SYSTEM FROM SEED TO TABLE

1 National Young Farmers Coalition Strategic Plan 2021–2026, 3,4.
2 Ingram, "Creating Credible Edibles," both on 127.
3 Brand, *Last Whole Earth Catalog*, 1.
4 Ibid.
5 Kauffman, *Hippie Food*, 188.
6 Musick, "The History of the Tilth Movement."
7 Hyde, L., interview with the author, September 23, 2015. All quotes from this interview.
8 Obituary of Charles S. Gould, *Portland* (Maine) *Press Herald*, December 8, 2013.
9 Barker, "History of Seed in the U.S."
10 Cooke, "Who Wants White Carrots?" 476; Barker, "History of Seed in the U.S."
11 In 1980, the US Supreme Court ruled 5–4 (in *Diamond v. Chakrabaty*) that living organisms could be patented.
12 Barker, "History of Seed in the U.S."
13 USDA/ERS, "Seed Industry Structure Is Characterized by Growth and Consolidation"; Barker, "History of Seed in the U.S."
14 Johnny's Selected Seeds, "History of Johnny's."
15 MOFGA, *Fertile Ground*, 23; Beem, "Rooted in the counterculture of the 1970s, Johnny's Selected Seeds."
16 Territorial Seeds, "About Us."
17 Whealy, *Gathering*, 23–24.
18 Ibid., 42.
19 Ibid., 44, 50.
20 Southern Exposure Seed Exchange, "Southern Exposure: The Early Years."
21 Yuan, "Why Slow Tools."
22 Rosen and McGrane, "The Revolution Will Not Be Catered"; Hoffman, *Steal This Book!* np.
23 Wickstrom, *The Food Conspiracy Cookbook*, 8; Curl, *For All the People*, 213.

24 Pollan, *The Omnivore's Dilemma*, 143.

25 Knupfer, *Food Co-ops in America*, 134.

26 Belasco, *Appetite for Change*, 141.

27 Ibid., 91; Schiferl and Boynton, "Comparative Performance Analysis," 338.

28 Pyle, "Farmers' Markets in the United States," 197.

29 Oakes, S. R. "Behind the Plate"; Sommer, *Farmers Markets in America*, 81.

30 Linstrom, "Farmer to Consumer Marketing"; Sommer, *Farmers Markets in America*, 38.

31 Sommer, *Farmers Markets in America*, 28; Sommer, Wing, and Aitkens, "Price Saving," 457.

32 Mackler, "Diesel fuel priority for farmers ended"; Kramer, "Truckers' Strike Intensifies in Violence and Disruption."

33 High-return crops could generate $3,000/acre then or $8,000/acre today.

34 *Mother Earth News* editors, "The Small Farm Plan by Booker T. Whatley."

35 Einseidler, telephone interview, September 9, 2015; Fox, interview with the author, November 3, 2015.

36 Barber, *The Third Plate*, 75.

37 Waters, *Coming to My Senses*, 149, 91.

38 Ibid., 111.

39 Ibid., 148.

40 Freedman and Meyer, *Ten Restaurants*, 391; McNamee, *Alice Waters and Chez Panisse*, 59, 82.

41 Freedman and Meyer, *Ten Restaurants*, both quotes on 369.

42 Bolois, "The 10 Dishes That Made My Career: Alice Waters"; Neff, "Bob Cannard."

43 Petersen, "Simple food, farm life and deliciousness."

44 Waters, *Coming to My Senses*, 183.

45 Hanchett, "US Tax Policy and the Shopping Center Boom," 1095.

46 Ibid., 1097, 1098.

47 Jackson, "National Land Use Policy Act."

48 Boehje, "Tax Policy and the Structure of Agriculture," 143.

49 Trustees of Reservations, "Trustees History"; Brewer, *Conservancy*, 38.

50 Georges River Land Trust, Georges River Land Trust 2018 Annual Report.

51 Kamp, *The United States of Arugula*, 142.

CHAPTER NINE: FREE THE PEOPLE, FREE THE LAND

1 Thoreau, "Civil Disobedience," 799–800.

2 USDA Press Release #0283.18.

3 USDA Study Team, *Report and Recommendations*, 1.

4 Ibid., 48.

5 Blobaum, "Inside Organic."

6 Gershuny, *Organic Revolutionary*, 98.

7 Youngberg and DeMuth, "Organic Agriculture in the United States," 314.

8 USDA, 2017 Agricultural Census.

9 USDA, *Summary Report on the Structure of Agriculture*, 11.

10 USDA National Commission on Small Farms, Section III: Introduction and Section II: Executive Summary.

11 KPMG, "A New Vision of Value."

12 Lee, "The True Cost of Food."

13 Herrington, interview with the author. All quotes from this interview unless otherwise noted.

14 Conway, *Get Back, Stay Back*, 179.

15 Goad, "Meet the 2019 Russell Libby Scholarship Winners."

16 von Tscharner Fleming, et al., 1.

17 Jenkins, Gordon, interview with the author, December 9, 2019.

18 Muhlke, "A Slow Food Festival Reaches Out."

19 Purdy, *This Land Is Our Land*, 28.

20 Ibid., xiii.

BIBLIOGRAPHY

Adams, Grace, and Edward Hutter. *The Mad Forties*. New York: Harper & Brothers Publishers, 1942.

Alcott, Amos Bronson. *Table-Talk*. Boston: Roberts Brothers, 1877.

Alcott, Amos Bronson, and Charles Lane. Letter excerpt in *The Dial*, July 10, 1843: 135–36.

———. "The Consociate Family Life," *New York Tribune* (September 1, 1843).

Alcott, Louisa May. *Transcendental Wild Oats and Excerpts from the Fruitlands Diary*. Introduction by William Henry Harrison. Carlisle, MA: Applewood Books, 1981.

Alex, Joel. Interview with the author, April 17, 2019.

Algiere, Jack. Lecture, Stone Barns Center Young Farmers Conference. Pocantico Hills, NY, December 5, 2018.

American Cheese Society. "About Us." Wayback Machine: cheesesociety.org, December 12, 1998.

"American Vegetarian Convention—A Detailed Report." New York City, May 1850. ivu.org.

Anderson, James. "'Fruitlands,' Historic Alcott Home Restored," *Table Talk: The American Authority upon Culinary Topics and Fashions of the Table*, Vol. 30, issue 12 (December 1915): 664–70.

Anson Mills. "Rustic Red Fife Bread Flour." ansonmills.com.

Atwater, Maxine. *Natural Foods Cookbook*. Concord, CA: Nitty Gritty Productions, 1972.

Barber, Dan. *The Third Plate: Field Notes on the Future of Food*. New York: Penguin Books, 2015.

Barker, Debbie. "History of Seed in the U.S.: The Untold American Revolution," Center for Food Safety, 2012. centerforfoodsafety.org.

Belasco, Warren. *Appetite for Change: How the Counterculture Took On the Food Industry*, 2nd edition. Ithaca, NY: Cornell University Press, 2nd edition, 2007.

———. "Food and the Counterculture: A Story of Bread and Politics," in *Food in Global History*, edited by Raymond Grew. New York: Routledge, 2018.

Beem, Edgar Allen. "Rooted in the counterculture of the 1970s, Johnny's Selected Seeds flourishing with the locavore movement," in *Down East Magazine* (March 2016): 38–41.

Berkowitz, Nancy, and Warren Berkowitz. Interview with the author, September 20, 2016.

Ibid., June 28, 2017.

Berry, Wendell. *The Unsettling of America: Culture and Agriculture* 2nd edition. San Francisco: Sierra Club, 1996.

Bidwell, Percy W. "The Agricultural Revolution in New England," in *American Historical Review*, Vol. 26, No. 4 (July 1921): 683–702.

Blobaum, Roger. "Inside Organic: USDA Had an Organic Farming Coordinator in 1980; Call for Reinstatement Now Made 30 Years Later," March 2010. rogerblobaum.com.

Bloodroot Collective. "About." bloodroot.com.

————. *The Political Palate: A Feminist Vegetarian Cookbook*. Bridgeport, CT: Sanguinaria Publishing, 1980.

Blum, Deborah. *The Poison Squad: One Chemist's Single-Minded Crusade for Food Safety at the Turn of the Twentieth Century*. New York: Penguin Books, 2018.

Bly, Chris, and John Bly. Interview with the author, October 15, 2015.

Bodwell, Joshua. "Year-Round Radicals," in *Maine Home + Design*, September 2007: 60–65.

Boehje, Michael. "Tax Policy and the Structure of Agriculture," in *Increasing Understanding of Public Problems and Policies* (1980), report of the 30th National Public Policy Education Conference. Vail, CO: Farm Foundation, 1980: 141–52.

Bolois, Justin. "The 10 Dishes That Made My Career: Alice Waters," *First We Feast* (April 20, 2015). firstwefeast.com.

Bourdain, Anthony. *See* Kanani.

Borsodi, Ralph. *This Ugly Civilization*. New York: Simon & Schuster, 1929.

Brand, Stewart, editor. *Last Whole Earth Catalog*. Menlo Park, CA: Portola Institute, 1971.

Brewer, Richard. *Conservancy: The Land Trust Movement in America*. Lebanon, NH: University Press of New England, 2003.

Bright, Jean Hay. *Meanwhile, Next Door to the Good Life: Homesteading in the 1970s in the shadows of Helen and Scott Nearing, and how it all—and they—ended up*. Rev. 10th-anniversary edition. Dixmont, ME: BrightBerry Press, 2013.

Brock, Alice May. *Alice's Restaurant Cookbook*. New York: Random House, 1969.

Brown, Dona. *Back to the Land: The Enduring Dream of Self-Sufficiency in Modern America*. Madison: University of Wisconsin Press, 2011.

Brown, James W. "Alimentary Discourse in Nineteenth-Century Social Theory, Pierre Leroux, Etienne Cabot, and Charles Fourier," in *Dalhousie French Studies*, Vol. 11 (Fall/Winter 1986): 72–96.

Brownson, Orestes A. *New Views of Christianity, Society, and the Church*. Boston: James Munroe and Company, 1836.

Buhs, Joshua Blu. "Scott Nearing as a Fortean (and Helen Knothe Nearing, too), *From an Oblique Angle* (April 21, 2015). joshuablubuhs.com.

Burnett, Peter. "State of the State Address." January 6, 1851. governors.library. ca.gov.

Carlson, Larry A. "The Inner Life of Fruitlands," in *Lives Out of Letters: Essays on American Literary Biography and Documentation in Honor of Robert N. Hudspeth*, edited by Robert D. Habich. Madison, NJ: Farleigh Dickinson University Press, 2004.

Carlyle, Thomas. *Shooting Niagara: And After?* London: Chapman and Hall, 1867.

———— and Ralph Waldo Emerson. *The Correspondence of Thomas Carlyle and Ralph Waldo Emerson, 1834–1872*. Vol. 1. Boston: James R. Osgood, 1883.

Carroll, Robert. "Guide to Home Cheese Making," in *Mother Earth News* (March/April 1986).

Channing, William Ellery. Letter to Thoreau, March 1845. walden.org.

Clopp, Doug. "New Roots Cooperative Farm: First New American Co-op Farm in Maine." Cooperative Development Institute, July 26, 2016.

Coleman, Eliot. Interview with the author, April 11, 2017.

Coleman, Melissa. *This Life Is in Your Hands: One Dream, Sixty Acres, and a Family's Heartbreak.* New York: HarperCollins, 2011.

Conrad, Randall. "Realizing Resistance: Thoreau and the First of August 1846, at Walden," the *Concord Saunterer, New Series,* Vol. 12/13 (2004/2005), 165–94.

Conway, Joseph. *Get Back, Stay Back: 2nd Generation Back-to-the-Landers in Maine.* Portland, ME: Leisure Labor, 2014.

Cooke, Kathy J. "Who Wants White Carrots? Congressional Seed Distribution, 1862 to 1923." *Journal of the Gilded Age and Progressive Era,* 17 (2018): 475–500.

Coyote, Peter. *Sleeping Where I Fall: A Chronicle.* Washington, DC: Counterpoint, 1998.

Culver, John C., and John Hyde, *American Dreamer: A Life of Henry A. Wallace.* New York: W. W. Norton, 2000.

Curl, John. *For All the People: Uncovering the Hidden History of Cooperation, Cooperative Movements, and Communalism in America.* Oakland, CA: PM Press, 2009.

Curtis, Olga. "Modern Kitchen Miracles," *Washington Post,* April 26, 1957: C2.

Cypress Grove Cheese, "Our Story." cypressgrovecheese.com.

Darlington, Jeanie. *Grow Your Own: An Introduction to Organic Gardening.* Berkeley: The Bookworks, 1971.

Davis, Jay. Interview with the author, September 22, 2015.

Dedmond, Francis B. "Burning the Root of the Hydra's Head: Thoreau and the Abolition Movement," *Concord Saunterer,* Vol. 11, No. 4 (Winter 1976): 1–8.

Delano, Sterling F. *Brook Farm: The Dark Side of Utopia.* Cambridge, MA: Belknap Press, 2004.

Diggers, The. "Let Me Live in a World Pure," The Digger Archives, Fall 1966. diggers.org.

———. "Men Are Out of Touch with the Earth," The Digger Archives, October 1967. diggers.org.

———. "Trip Without a Ticket," The Digger Archives, October 1966. diggers.org.

Dragonwagon, Crescent. *The Commune Cookbook.* New York: Simon & Schuster, 1972.

Edwards, Eliza Harvey. "Arden: The Architecture and Planning of a Delaware Utopia." Unpublished master's thesis, University of Pennsylvania, 1993.

Einseidler, Diane. Telephone interview with the author, September 9, 2015.

Elzey, Lydia. "An Elementary School Garden Project," in *American Biology Teacher,* Vol. 6, No. 5 (February 1944): 115–18.

Emerson, Ralph Waldo. "Concord Hymn," 1837. poetryfoundation.org.

———. *Journals and Miscellaneous Notebooks of Ralph Waldo Emerson,* edited by William H. Gilman, Ralph H. Orth, et al., 16 vols., Cambridge, MA: Harvard University Press.

———. "Sermon CIV," in *The Complete Sermons of Ralph Waldo Emerson,* Vol. 3, edited by Ronald A. Bosco. Columbia: University of Missouri Press, 1991.

Epstein, Abraham. "Have American Wages Permitted an American Standard of Living? A Review of the Important Inquiries and Their Findings, 1890–1920," in *Annals of the American Academy of Political and Social Science,* Vol. 97 (September 1921): 169–90.

Evans, Pat. "All 50 States Ranked for Beer," in *Beer Connoisseur*, March 4, 2019.

Fabian, Vera. Telephone interview with the author, December 11, 2018.

———. Interview with the author, December 9, 2019.

Fairchild, Richard. *The Modern Utopian: Alternative Communities of the '60s and '70s.* Port Townsend, WA: Process Media Inc., 2010.

"Farm Workers Press Lettuce Boycott," in the *Harvard Crimson* (November 13, 1970).

Forman, Alex. "San Francisco style: The diggers and the love revolution," in *Anarchy 77*, Vol. 7, No. 7 (July 1967).

Fourier, Charles. *The Theory of the Four Movements*, edited by Gareth Stedman Jones and Ian Patterson. New York: Cambridge University Press, 1996.

Fox, Doug. Interview with the author, November 3, 2015.

Frame, Caitlin. Interview with the author, May 25, 2017.

———. Lecture, "Real Organic Project Symposium." Dartmouth College, Hanover, NH, March 2, 2019.

Frame, Caitlin, and Andy Smith. "Know Your Farmer," Real Organic Project. realorganicproject.org.

Freedman, Paul, and Danny Meyer. *Ten Restaurants That Changed America.* New York: Liveright Publishing, 2016.

Frisch, Tracy. "Taking on Food Justice with Soul Fire Farm's Leah Penniman," in *Eco Farming Daily*, April 2017.

———. "To Free Ourselves, We Must Feed Ourselves," *The Sun* (July 2019).

"Fruitlands." alcott.net

George, Henry. *Progress and Poverty: An Inquiry into the Cause of Industrial Depressions and of Increase of Want with Increase of Wealth: The Remedy*, 50th-anniversary edition. New York: Robert Schalkenbach Foundation, 1935.

Georges River Land Trust. Georges River Land Trust 2018 Annual Report.

Gershuny, Grace. *Organic Revolutionary: A Memoir of the Movement for Real Food, Planetary Healing, and Human Liberation.* Vermont: Joe's Brook Press, 2016.

Ginsberg, Allen. "Footnote to 'Howl.'" poetryfoundation.org.

Goad, Meredith. "Grain partnership between Maine brewers, farmers in peril," *Portland Press Herald* (March 28, 2014).

———. "Meet the 2019 Russell Libby Scholarship Winners," *Portland Press Herald* (April 21, 2019).

Goldman, M. C., and William H. Hylton, eds. *The Basic Book of Organically Grown Foods.* Emmaus, PA: Rodale Press, 1972.

Goodman, Jim. "With the USDA's Blessing, CAFOs Are Driving Organic Dairy Farmers Out of Business." *Rural American: In These Times* (February 23, 2018).

Graham, Sylvester. *A Treatise on Bread, and Bread-Making.* Boston: Light & Stearns, 1837.

Gray, Erik. "Radical Teaching: Scott and Helen Nearing's Impact on Maine's Natural Food Revival," *Maine History* 48, 2 (2014): 267–84.

Gregory, David. "Karl Marx's and Friedrich Engels' Knowledge of French Socialism in 1842–43," *Historical Reflections/Réflexions Historiques*, Vol. 10, No. 1 (Spring 1983): 143–93.

Grogan, Emmett. *Ringolevio: A Life Played for Keeps.* New York: New York Review of Books, 2008 (Boston: Little, Brown and Co., 1972).

Gross, Robert A. "Culture and Cultivation: Agriculture and Society in Thoreau's Concord," *Journal of American History*, Vol. 69, No. 1 (June 1982): 42–61.

Guarneri, Carl J. "The Americanization of Utopia: Fourierism and the Dilemma of Utopian Dissent in the United States," *Utopian Studies*, Vol. 5., No. 1 (1994): 72–88.

———. "Reconstructing the Antebellum Communitarian Movement: Oneida and Fourierism," *Journal of the Early Republic*, Vol. 16, No. 3 (Autumn, 1996): 463–88.

Gunderson, Gordon W. "The National School Lunch Program: Background and Development." US Government Printing Office, 0-429-783, 1971.

Halper, Emmanuel B. "Evolution and Maturity of the American Supermarket During World War II," *Real Property, Probate and Trust Journal*, Vol. 41, No. 2 (Summer 2006): 253–356.

Hamilton, Neil D. "America's New Agrarians: Policy Opportunities and Legal Innovations to Support New Farmers," *Fordham Environmental Law Review*, Vol. 22, No. 3 (Fall 2011): 523–62.

Hamilton, Shane. "The Economies and Conveniences of Modern-Day Living: Frozen Foods and Mass Marketing, 1945 65," *Business History Review*, Vol 77, No. 1 (Spring 2003): 33–60.

Hanchett, Thomas W. "U.S. Tax Policy and the Shopping Center Boom of the 1950s and 1960s," *American Historical Review*, Vol. 101, No. 4 (October 1996): 1082–1110.

Hannaford, Kathryn. *Cosmic Cookery*. Berkeley: Starmast Publications, 1974.

Harding, Walter. "The First Year Sales of Thoreau's *Walden*," *Thoreau Society Bulletin*, No. 117 (Fall 1971): 1–3.

Harlem Valley Homestead, "Down the Road." harlemvalleyhomestead.com.

Hartman, Stephanie. "The Political Palate: Reading Commune Cookbooks," *Gastronomica*, Vol. 3, No. 2 (Spring 2003): 29–40.

Harvey, Eric. Interview with the author, December 5, 2018.

Harvey, Nelson. "Why Are Young, Educated Americans Going Back to the Farm?" *Huffington Post* (October 2, 2011).

Hawthorne, Nathaniel. Journal entry, September 1, 1842.

Heart of Maine website, accessed a page from June 14, 2007 via the Wayback machine.

Heffernan, Ashley. "Farm store brings fresh produce back to Chicora-Cherokee neighborhood," *Charleston Regional Business Journal* (June 7, 2016).

Herrington, Kelsey. Interview with the author, March 28, 2017.

Heuchling, Fred G. "Children's Gardens in Chicago," *The American Biology Teacher*, Vol. 6, No. 5, (February 1944): 103–107.

Hoffman, Abbie. *Steal This Book*. Pirate edition, 1971.

Holmes, Ben. Interview with the author, June 7, 2017.

Horton, Lucy. *Country Commune Cooking*. New York: Coward, McGann & Geoghegan, 1972.

Houriet, Robert. *Getting Back Together*. London: Abacus Books, 1973.

———. "Life and Death of a Commune Called Oz," *New York Times* (February 16, 1969): Section SM, 30+ff.

Hoyt, Sherm. Interview with the author, September 16, 2015.

Hurt, R. Douglas. *American Agriculture: A Brief History*. Ames: Iowa State Press, 1994.

Hyde, Les. Interview with the author, September 23, 2015.

Ingram, Mrill, and Helen Ingram. "Creating Credible Edibles: The Organic Agriculture Movement and the Emergence of U.S. Federal Organic Standards." In *Routing the Opposition: Social Movements, Public Policy, and Democracy*, edited by David S. Meyer, Valerie Jenness, and Helen Ingram. Minneapolis: University of Minnesota Press, 2005.

Jackson, Henry (Senator). "National Land Use Policy Act," S. 3354, 91st Congress, 2nd Session (1970).

Jacob, Jeffrey. *New Pioneers: The Back-to-the-Land Movement and the Search for a Sustainable Future*. College Station: Pennsylvania State University Press, 1997.

James, P., M. C. Arcaya, D. M. Parker, R. D. Tucker-Seeley, and S. V. Subramanian. "Do minority and poor neighborhoods have higher access to fast-food restaurants in the United States?" *Health Place*, Vol. 29, 2014: 10–17.

Jenkins, Germaine. In "Nom, Nom, Nom," special insert in *Skirt* magazine. November 2016, 44.

Jenkins, Gordon. Interview with the author, December 9, 2019.

Johnny's Selected Seeds. "History of Johnny's." johnnyseeds.com.

Johnson, Linnea. "Nearing Enough: Simple Living Lessons, *Mother Earth News*, October 2003.

Kamp, David. *The United States of Arugula: The Sun Dried, Cold Pressed, Dark Roasted, Extra Virgin Story of the American Food Revolution*. New York: Broadway Books, 2006.

Kanani, Rahim. "The World According to Anthony Bourdain," *Food and Wine*, May 24, 2017.

Kauffman, Jonathan. *Hippie Food: How Back-to-the-landers, Longhairs, and Revolutionaries Changed the Way We Eat*. New York: William Morrow, 2018.

Kaymen, Samuel, "Oral History," audio recording, December 1998. NOFA Archive. Special Collections and University Archives, University of Massachusetts Amherst.

Killinger, Margaret O'Neal. *Helen Knothe Nearing: A Biography*. PhD dissertation. University of Maine, 2004.

———. *The Good Life of Helen K. Nearing*. Lebanon, NH: University Press of New England, 2007.

Knupfer, Anne Meis. *Food Co-ops in America: Communities, Consumption, and Economic Democracy*. Ithaca, NY: Cornell University Press, 2013.

KPMG International. "A New Vision of Value: Connecting corporate and societal value creation." 2014.

Kramer, Larry. "Truckers' Strike Intensifies in Violence and Disruption," *Washington Post* (June 22, 1979).

Kritzberg, Barry. "Thoreau, Slavery, and Resistance to Civil Government, *Massachusetts Review*, Vol. 30, No. 4 (Winter 1989): 535–65.

Laura Chenel, "40 Years of Love (and Goat Milk)." laurachenel.com.

Lee, Mike. "The True Cost of Food," *Medium* (December 18, 2016).

Legnini, Jenn. Interview with the author, December 7, 2017.

Lemire, Elise. *Black Walden: Slavery and Its Aftermath in Concord, Massachusetts*. Philadelphia: University of Pennsylvania Press, 2009.

Leopold, Aldo. *A Sand County Almanac and Sketches Here and There*. New York: Oxford University Press, 1949.

Levenstein, Harvey. *Fear of Food: A History of Why We Worry about What We Eat*. Chicago: University of Chicago Press, 2013.

———. *Revolution at the Table: The Transformation of the American Diet*. Berkeley: University of California Press, 2003.

Levi, Jane E. *Food in Utopia: Eating Our Way to Perfection*. Dissertation, King's College London, 2014.

Levine, Susan. *School Lunch Politics: The Surprising History of America's Favorite Welfare Program*. Princeton, NJ: Princeton University Press, 2010.

Linstrom, H. R. "Farmer to Consumer Marketing," Economics, Statistics, and Cooperative Service, U.S. Department of Agriculture, ESCS-01 (February 1978).

Lozano, Gabby. "24 LGBTQ+ Farms and Organizations Celebrating Community through Food and Agriculture." foodtank.com/news/2020/06/16.

"Lucy Horton Interview," transcript of audio recording. June 25, 1996. Communal Studies Gallery, Oral History Collection, University of Southern Indiana, Evansville.

Mackler, "Diesel fuel priority for farmers ended," *Huron Daily Plainsman* (June 22, 1979), 1.

Magliari, Michael F. "Free State Slavery: Bound Indian Labor and Slave Trafficking in California's Sacramento Valley, 1850–1864," *Pacific Historical Review*, Vol. 81, No. 2 (May 2012): 155–92.

"Maine Farm and Seafood Product Directory," University of Maine Cooperative Extension. extension.umaine.edu.

Maine Organic Farmers and Gardeners Association. *Fertile Ground: Celebrating 40 Years of MOFGA*. Newcastle, ME: Lincoln County Publishing, 2011.

Martel, Craig. Interview with the author, April 13, 2017.

Marx, Karl. *Capital: A Critique of Political Economy*, Vol. 1. New York: Vintage, 1970 reprint.

Mason, W. E. "Food Adulterations," *North American Review*, Vol. 170, No. 521 (April 1900), 548–52.

Matteson, John. *Eden's Outcasts: The Story of Louisa May Alcott and Her Father*. New York: W. W. Norton, 2008.

Matthews, James W. "An Early Brook Farm Letter," *New England Quarterly*, Vol. 53, No. 2 (June 1980): 226–30.

Maynard, W. Barksdale. *Walden Pond: A History*. New York: Oxford University Press, 2004.

McKinney, Connee Wright. "What the Arden School Can Teach Us: Hard Lessons in Community Building," master's thesis, University of Delaware, 2004.

McNamee, Thomas. *Alice Waters and Chez Panisse: The Romantic, Impractical, Often Eccentric, Ultimately Brilliant Making of a Food Revolution*. New York: Penguin Books, 2008.

Miller, Timothy. *The Quest for Utopia in Twentieth-Century America, Volume I: 1900–1960*. Syracuse, NY: Syracuse University Press, 1998.

More, Sir Thomas. *Utopia*. Edited by Edward Surtz, SJ. New Haven, CT: Yale University Press, 1964.

Morris, William. Preface to "The Nature of Gothic." Kelmscott Press, 1892.

Moss, Michael. *Salt, Sugar, Fat: How the Food Giants Hooked Us.* New York: Random House, Trade Paperback Edition, 2014.

Mother Earth News editors. "Dr. Ralph Borsodi: Legendary Back-to-the-Land Figure," *Mother Earth News*, March/April 1974.

———. "The Small Farm Plan by Booker T. Whatley," *Mother Earth News*, May/June 1982.

Mudge, Sam. Interview with the author, March 29, 2016.

Muhlke, Christine. "The Hippies Have Won," *New York Times* (April 4, 2017).

———. "A Slow Food Festival Reaches Out to the Uncommitted," *New York Times* (September 2, 2008).

Murray's Cheese Blog, "A History of Loving Cheese in the USA," January 20, 2017. blog.murrayscheese.com.

Musick, Mark. "The History of the Tilth Movement," Tilth Alliance. tilthalliance.org.

Nader, Ralph. "Watch That Hamburger," *New Republic* (August 19, 1967): 15–16.

National Young Farmers Coalition. "Building a Future with Farmers II: Results and Recommendations from the National Young Farmer Survey," November 2017.

———. "National Young Farmers Coalition Strategic Plan 2021–2026," October 22, 2021.

Nearing, Helen (Helen Knothe). Interview with Nina Ellis, NPR, November 1989. Recording housed at the Maine Folklife Center Archive and the University of Maine at Orono.

———. *Loving and Leaving the Good Life.* White River Junction, VT: Chelsea Green Publishing, 1992.

———. *Our Home Made of Stone: Building in Our Seventies and Nineties.* Rockport, ME: Down East Books, 1983.

———. *Simple Food for the Good Life: Random Acts of Cooking and Pithy Quotations.* White River Junction, VT: Chelsea Green Publishing, 1990.

———. "Thoreau Judged in His Own Time," undated, unpublished manuscript, housed at Walden Woods Project Archive.

———. Untitled public lecture, University of Iowa, March 25, 1980. Recording housed in the Maine Folklife Center Archive, University of Maine at Orono.

Nearing, Helen and Scott Nearing. *The Good Life.* New York: Schocken Books, 1970.

———. "Raising Children, Homesteading Books, and Other Wisdom from Helen and Scott Nearing," *Mother Earth News*, November/December 1978.

———. "Summing Up," *Down East Magazine*, July 1979: 58–63.

Nearing, Scott. "The Adequacy of American Wages," *The Annals of the American Academy of Political and Social Science*, Vol. 59, *The American Industrial Opportunity* (May 1915): 111–124.

———. "Arden Town," unpublished manuscript, 1913. Housed at Walden Woods Project Archive.

———. *Black America.* New York: The Vanguard Press, 1929.

———. Letter to Helen. January 18, 1944. Housed at Walden Woods Project Archive.

———. *The Making of a Radical: A Political Autobiography.* White River Junction,

VT: Chelsea Green Publishing, 2000.

———. *Reducing the Cost of Living*. Philadelphia: G. W. Jacobs & Co., 1914.

———. "The Relation of Simple Living to Social Radicalism," undated typescript, Series 1.310.a Housed at Walden Woods Project Archive.

———. "Scott Nearing on The Good Life," *YouTube*.

———. Untitled tribute to Frank Stephens, *Frank Stephens' Songs and Tributes from Old Friends*, 1959. Centennial Booklet #1, 1959.

Nearing, Scott and Nellie Seeds. *Woman and Social Progress*. New York: Macmillan, 1912.

Neff, Kirsten Jones. "Bob Cannard: Visionary Farmer and Educator and His Green String Farm," *Edible Marin & Wine Country* (June 1, 2013).

Neuhaus, Jessamyn. "The Way to a Man's Heart: Gender Roles, Domestic Ideology, and Cookbooks in the 1950s," *Journal of Social History*, Vol. 32, No. 3 (Spring 1999): 529–55.

New England Cheese Making Supply Company, "Our Story." cheesemaking.com.

Newhall, Priscilla. "Arden—A Colony of Pleasure and Profit: Little Settlement in a Wood Overlooking the Delaware River, Where Live a Company of Thinkers and Workers," *Suburban Life*, Vol. 13 (August 1911): 75–76.

Nosowitz, Dan. "The Real Organic Project: Disgusted with the USDA, Farmers Make Their Own Organic Label," *Modern Farmer* (March 5, 2018).

Noyes, Tristan. Interview with the author, May 23, 2017.

Oakes, Kaya. *Slanted and Enchanted: The Evolution of Indie Culture*. New York: Holt Paperbacks, 2009.

Oakes, Summer Rayne. "Behind the Plate with Greenmarket Founder Barry Benepe and Foodstand," *Medium* (May 24, 2016).

O'Sullivan, Robin. *American Organic: A Cultural History of Farming, Gardening, Shopping, and Eating*. Lawrence: University Press of Kansas, 2015.

Patel, Raj. "Survival Pending Revolution: What the Black Panthers Can Teach the US Food Movement," *Food First Backgrounder*, Vol. 18, No. 2 (Summer 2012): 1–3.

Penniman, Leah. "4 Not-So-Easy Ways to Dismantle Racism in the Food System," *YES! Magazine* (April 27, 2017).

———. *Farming While Black*. White River Junction, VT: Chelsea Green Publishing, 2018.

———. "A New Generation of Black Farmers Is Returning to the Land," *Civil Eats* (November 27, 2019).

———. Interview with the author. February 15, 2019.

——— and Blain Snipstal. "Chapter Three: Regeneration," *Land Justice: Reimagining Land, Food, and the Commons in the United States*. Edited by Justine M. Williams and Eric Holt-Giménez. Oakland, CA: Food First Books, 2017.

Perry, Charles. *The Haight-Ashbury: A History*. New York: Random House, 1984.

Petersen, Laura. "Simple food, farm life and deliciousness: Chef Alan Tangren's life comes full circle," *The Union* (August 29, 2017).

Petrulionis, Sandra Harbart. "'Swelling That Great Tide of Humanity': The Concord, Massachusetts, Female Anti-Slavery Society," *New England Quarterly*, Vol. 74, No. 3 (September 2001): 385–418.

Petty, Adrienne, and Mark Schultz. "African-American Farmers and the USDA: 150 Years of Discrimination," *Agricultural History*, Vol. 87, No. 3 (Summer 2013): 332–43.

Phillips, Kevin. *Wealth and Democracy: How Great Fortunes and Government Created America's Aristocracy.* New York: Broadway Books, 2002.

Poches, Olivia. "The American Chamber of Horrors," *Go Big Red,* May 24, 2019. gobigread.wisc.edu.

Pollan, Michael. *The Omnivore's Dilemma: A Natural History of Four Meals.* New York: Penguin Press, 2006.

Portland Press Herald (Portland, Maine). Obituary of Charlie Gould, December 8, 2013.

Purdy, Jedidiah. *A Tolerable Anarchy: Rebels, Reactionaries, and the Making of American Freedom.* New York: Vintage, 2010.

———. *This Land Is Our Land: The Struggle for a New Commonwealth.* Princeton, NJ: Princeton University Press, 2019.

Pyle, Jane. "Farmers' Markets in the United States: Functional Anachronisms," *The Geographical Review,* Vol. LXI, No. 2 (April 1971): 167–97.

"Question of Slavery, The," *The Harbinger,* June 21, 1845: 29–31.

"Radical professor recalls Arden life of 50 years ago," *Morning News,* Wilmington, DE (May 8, 1973): 9.

Reed, Amy L., ed. *Letters from Brook Farm, 1844–1847.* Poughkeepsie, NY: Vassar College, 1928.

Richmond, Todd. "Ag secretary: No guarantee small dairy farms will survive," *Washington Post* (October 1, 2019).

Ridley, Kim. "Back to the Land, 2.0," *Down East Magazine,* September 2013: 100–111.

Rodabaugh, W. J. "A Nation of Sots: When drinking was a patriotic duty," *New Republic* (September 29, 1979).

Rodale, J. I. *Pay Dirt: Farming & Gardening with Composts.* New York: Devin-Adair Co., 1945.

Rohman, Keith, "Let Us Boycott Lettuce," *Miscellany News,* Vol. LXV, No. 10 (November 12, 1972).

Rooney, Ben. Interview with the author, March 20, 2017.

Rosen, Laurel, and Sally McGrane. "The Revolution Will Not Be Catered: How Bay Area food collectives of the '60s set the stage for today's sophisticated food tastes," *SFGate,* March 8, 2000.

Rosenvall, Vernice, Mabel H. Miller, and Dora Flack. *The Classic Wheat for Man Cookbook.* Santa Barbara, CA: Woodbridge Press, 1975.

Sage, Tyler. Interview with the author, December 17, 2018.

———. Telephone conversation with the author, September 12, 2015.

Saloutos, Theodore. "The Agricultural Problem and Nineteenth-Century Industrialism," *Agricultural History,* Vol. 22, No. 3 (July 1948): 156–74.

Sanborn, F. B., editor. *Bronson Alcott at Alcott House, England, and Fruitlands, New England (1842–44).* Cedar Rapids, IA: The Torch Press, 1908.

——— and William T. Harris. *A. Bronson Alcott: His Life and Philosophy,* Vol. II. Boston: Roberts Brothers, 1893.

Sargent, Lyman Tower. "The Social and Political Ideas of the American Communitarians: A Comparison of Religious and Secular Communes Founded Before 1850." *Utopian Studies,* 1991, No. 3 (1991): 37–58.

———. "The Three Faces of Utopianism Revisited," *Utopian Studies* 5:1 (1994): 1–38.

Scharnhorst, Gary. "Five Uncollected Contemporary Reviews of 'Walden.'" *The*

Thoreau Society Bulletin, No. 160 (Summer 1982): 1–3.

Schiferl, Elizabeth A., and Robert D. Boynton. "A Comparative Performance Analysis of New Wave Food Cooperatives and Private Food Stores," *Journal of Consumer Affairs* (Winter 1983): 336–53.

Schreiner, Samuel. *The Concord Quartet: Alcott, Emerson, Hawthorne, Thoreau and the Friendship That Freed the American Mind*. Hoboken, NJ: Wiley & Sons, 2006.

Sears, Clara Endicott, and Louisa May Alcott. *Bronson Alcott's Fruitlands*. Boston: Houghton Mifflin, 1915.

Shapiro, Laura. *Something from the Oven: Reinventing Dinner in 1950s America*. New York: Penguin Books, 2005.

Shepard, Odell. *The Journals of Bronson Alcott*. Boston: Little, Brown and Co., 1938.

Sherman, General William T. "Special Field Orders No. 15," in the Field, Savannah, GA, January 16, 1865.

Siegel, Elizabeth. Interview with the author, April 17, 2017.

Sinclair, Upton. *The Jungle*. New York, Doubleday, Page & Co., 1906.

Smith, Andy. Interview with the author, May 25, 2017.

Smith, Peter. "Brotherhood Spirit in Flesh Soup, or a Recipe Calling for Love," *Smithsonian Magazine*. February 13, 2012.

Smolensky, Eugene, and Robert Plotnick. "Inequality and Poverty in the United States: 1900 to 1990" (Discussion Paper #998-93). Madison, WI: Institution for Research on Poverty, March 1993.

Sokolov, Raymond A. "The Food at the Heart of Commune Life," *New York Times* (December 2, 1971).

Sommer, Robert. *Farmers Markets of America: A Renaissance*. Santa Barbara, CA: Capra Press, 1980.

———, Margaret Wing, and Susan Aitkens. "Price Savings to Consumers at Farmers' Markets," *The Journal of Consumer Affairs*, Vol. 14, No. 2 (Winter 1980): 452–62.

Southern Exposure Seed Exchange. "Southern Exposure: The Early Years." southernexposure.com.

Starkman, Naomi. "The Radical Necessity of Cooking: Mollie Katzen, Vegetablist," *Civil Eats* (March 18, 2010).

Stephens, Frank. "The Arden Enclave," *Enclaves of Economic Rent*, Vol. 4. Edited by Charles White Huntington. Harvard, MA: Fisk Warren, 1924.

Stickley, Gustave. *More Craftsmen Homes*. New York: Craftsman Publishing Company, 1912.

Stir the Pots Blog, April 1, 2020. stirthepots.com.

Stoller, Leo. "Thoreau's Doctrine of Simplicity," *New England Quarterly*, Vol. 29, No. 4 (December 1956): 443–61.

Taylor, Mark. "Utopia by Taxation: Frank Stephens and the Single Tax Community of Arden, Delaware," *Pennsylvania Magazine of History and Biography*, Vol. 126, No. 2 (April 2002): 305–325.

Territorial Seed Company. "About Us." territorialseed.com.

Thomas, George E. *William L. Price: Arts and Crafts to Modern Design*. New York: Princeton Architectural Press, 2000.

Thoreau, Henry David. "Civil Disobedience," rep. in *The Selected Works of Thoreau*,

Cambridge Edition. Revised and with a new introduction by Walter Harding. Boston: Houghton Mifflin, 1975.

———. "Commencement Address." In *Familiar Letters of Henry David Thoreau*, edited by F. B. Sanborn. Boston: Houghton, Mifflin & Co, 1894.

———. *Journal I*, March 3, 1841. walden.org.

———. *Journal III*, March 2, 1852. walden.org.

———. *Journal VIII*, November 5, 1855. walden.org.

———. *Journal XII*, April 23, 1859. walden.org.

———. "Slavery in Massachusetts." July 4, 1854. walden.org.

———. *Walden*, rep. in *The Selected Works of Thoreau, Cambridge Edition*. Revised and with a new introduction by Walter Harding. Boston: Houghton Mifflin, 1975.

True Light Beavers. *Eat, Fast, Feast: A Tribal Cookbook by the True Light Beavers*. Garden City, NY: Doubleday & Company, Inc., 1972.

Trunzo, Sara. Interview with the author, September 22, 2015.

Trustees of Reservations. "Trustees History." thetrustees.org.

Turner, Katherine Leonard. *How the Other Half Ate: A History of Working-Class Meals in the Turn of the Century*. Berkeley: University of California Press, 2014.

United Nations Study Group on The Economics of Ecosystems and Biodiversity (TEEB), "Measuring What Matters in Agriculture and Food Systems: A synthesis of the results and recommendations of TEEB for Agriculture and Food's Scientific and Economic Foundations Report." Geneva: UN Environment, 2018.

US Bureau of Labor. "City Gardens in Wartime," *Monthly Labor Review*. Vol. 61, No. 4 (October 1945): 644–50.

———. "Retail Price of Food, 1890–1906," *Bulletin of the United States Bureau of Labor*, Vol XV, Nos. 1–100 (July 1907): 175–328.

US Department of Agriculture, *2012 Census of Agriculture*. Washington, DC: National Agricultural Statistics Service, 2014.

———. *2017 Census of Agriculture*. Washington, DC: National Agricultural Statistics Service, 2019.

———. "Press Release #0283.18: Secretary Perdue Details Functions in the Event of a Lapse in Federal Funds." Washington, DC: December 21, 2018.

———. *Summary Report on the Structure of Agriculture*. With a foreword by Secretary of Agriculture Bob Berglund, "A Time to Choose." Washington, DC: January 1981.

US Department of Agriculture, Economic Research Service, "Seed Industry Structure Is Characterized by Growth and Consolidation," AIB-786. 25–29. US Department of Agriculture, National Commission on Small Farms. *A Time to Act: A Report of the USDA National Commission on Small Farms*. Washington, DC: January 1998.

US Department of Agriculture Study Team on Organic Farming. *Report and Recommendations on Organic Farming*. Washington, DC: July 1980.

US Food and Drug Administration. "80 Years of the Federal Food, Drug, and Cosmetic Act," *Virtual Exhibits of FDA History*. fda.gov.

Van Gelder, Sarah. "'Slow Food' Pioneer's Love for Food Ripened into a Life's Work," *Our World*, January 6, 2014. ourworld.unu.edu.

Vespucci, Amerigo. "First Voyage of Amerigo Vespucci." Project Gutenberg.

von Tscharner Fleming, Severine, Rivera Sun, Charlie Macquarie, Patrick Kiley, Marada Cook, Abby Sadauckus, Lance Lee, and John Patrick. *Mainifesta: An Un-monograph Celebrating the Maiden Voyage of Maine Sail Freight*. Creative Commons non-commercial, 2015.

Walker, Jesse. "Delaware's Odd, Beautiful, Contentious, Private Utopia," *Reason*, November 2017. reason.com/2017/10/14.

Wallach, Jennifer, ed. *American Appetites: A Documentary Reader*. Fayetteville: University of Arkansas Press, 2014.

———. *How America Eats: A Social History of U. S. Food and Culture*. Lanham, MD: Rowman & Littlefield, 2014.

Walls, Laura. *Henry David Thoreau: A Life*. Chicago: University of Chicago Press, 2017.

Walton, Levi. Interview with the author, February 25, 2016.

Waters, Alice. *Coming to My Senses: The Making of a Counterculture Cook*. New York: Clarkson Potter, 2017.

Weller, Sheila. "Suddenly That Summer," *Vanity Fair* (July 2012).

Whatley, Booker T. *How to Make $100,000 Farming 25 Acres*. Emmaus, PA: Regenerative Agriculture Association of the Rodale Institute, 1987.

Whealy, Diane Ott. *Gathering: Memoir of a Seed Saver*. Decorah, IA: Seed Saver Exchange, 2011.

White, Nora E. "Farming in the time of pandemic: Small farms demonstrate flexibility, innovation, and hope," *Journal of Agriculture, Food Systems, and Community Development*, March 19, 2021.

Whitfield, Stephen J. *Scott Nearing: Apostle of American Radicalism*. New York: Columbia University Press, 1974.

Wickstrom, Lois. *The Food Conspiracy Cookbook: How to Start a Neighborhood Buying Club and Eating Cheaply*. San Francisco: 101 Productions, 1974.

Wiencek, Henry. "A Delaware Delight: The Oasis Called Arden," *Smithsonian*, 23.2 (May 1992): 124–42.

Wilgus, Craig. Interview with the author, November 11, 2015.

Wu, Tim. "That Flour You Bought Could Be the Future of the U.S. Economy," *New York Times*, July 24, 2020.

Young, James Harvey. "The Pig That Fell into the Privy: Upton Sinclair's 'The Jungle' and the Meat Inspection Amendments of 1906," *Bulletin of the History of Medicine*, Vol. 59, No. 4 (Winter 1985): 467–80.

Young, Kelby, and Pam Young. Interview with the author, April 5, 2017.

Youngberg, Garth, and Suzanne P. DeMuth. "Organic agriculture in the United States: a 30-year retrospective," *Renewable Agriculture and Food Systems*, Vol. 28, Issue 4 (December 2013): 294–328.

Yuan, Linyee. "Why Slow Tools for Young Farmers Can Create a New Food Future," *Mold*, March 22, 2016. thisismold.com.

Ziegelman, Jane, and Andrew Coe. *A Square Meal: A Culinary History of the Great Depression*. New York: HarperCollins, 2016.

Zonderman, David A. "George Ripley's Unpublished Lecture on Charles Fourier," *Studies in the American Renaissance* (1982): 185–208.

A NOTE ABOUT THE AUTHOR

Margot Anne Kelley holds a PhD in American literature and an MFA in media and performing arts. She taught at the college level for almost twenty-five years. Since leaving academia, she served as the editor of the literary journal the *Maine Review* and cofounded a community-development corporation that runs a food pantry and community garden, among other projects. Like *Foodtopia*, most of her books, essays, and collaborative art projects explore the diverse relationships people have with the natural world.

A NOTE ON THE TYPE

Foodtopia has been set in Goudy Old Style. Designed by Frederic W. Goudy for the American Type Founders in 1915, the old-style serif typeface takes inspiration from printing during the Italian Renaissance. The diamond shape of the dots on the *i*, *j*, and punctuation points give the sturdy typeface an eccentric touch. Goudy was perhaps the best-known and most prolific type designer of his era: by the time he passed away in 1947, Goudy had designed 122 typefaces.

Design & composition by Tammy Ackerman